MOLECULAR BIOLOGY
INTELLIGENCE
UNIT

Intermediate Filaments

Jesús M. Paramio, Ph.D.
Department of Molecular and Cell Biology
CIEMAT
Madrid, Spain

LANDES BIOSCIENCE / EUREKAH.COM
GEORGETOWN, TEXAS
U.S.A.

SPRINGER SCIENCE+BUSINESS MEDIA
NEW YORK, NEW YORK
U.S.A.

INTERMEDIATE FILAMENTS

Molecular Biology Intelligence Unit

Landes Bioscience / Eurekah.com
Springer Science+Business Media, LLC

ISBN: 0-387-33780-6   Printed on acid-free paper.

Springer Science+Business Media, LLC, 233 Spring Street, New York, New York 10013, U.S.A.
http://www.springer.com

Please address all inquiries to the Publishers:
Landes Bioscience / Eurekah.com, 810 South Church Street, Georgetown, Texas 78626, U.S.A.
Phone: 512/ 863 7762; FAX: 512/ 863 0081
http://www.eurekah.com
http://www.landesbioscience.com

Printed in the United States of America.

9 8 7 6 5 4 3 2 1

Library of Congress Cataloging-in-Publication Data

Intermediate filaments / [edited by] Jesús M. Paramio.
    p. ; cm. -- (Molecular biology intelligence unit)
  Includes bibliographical references and index.
  ISBN 0-387-33780-6 (alk. paper)
    1. Cytoplasmic filaments.   I. Paramio, Jesús M. II. Series:
  Molecular biology intelligence unit (Unnumbered)
  [DNLM: 1. Intermediate Filament Proteins. 2. Gene Expression.
  3. Keratin--physiology.   QU 55 I609 2006]
  QH603.C95I58 2006
  611'.0181--dc22

                              2006009316

CONTENTS

EDITOR

Jesús M. Paramio
Department of Molecular and Cell Biology
CIEMAT
Madrid, Spain
Email: jesusm.paramio@ciemat.es
Chapter 10

CONTRIBUTORS

Miroslav Blumenberg
The Departments of Dermatology
   and Biochemistry
and
The Cancer Institute
NYU School of Medicine
New York, New York, U.S.A.
Email: blumem01@ med.nyu.edu
Chapter 7

Ana Bravo
Department of Animal Pathology
Veterinary School of Lugo
University of Santiago de Compostela
Madrid, Spain
Chapter 8

M. Llanos Casanova
Department of Molecular
   and Cell Biology
CIEMAT
Madrid, Spain
Email: llanos.casanova@ciemat.es
Chapter 8

Gee Y. Ching
Departments of Pathology and Anatomy
   and Cell Biology
Columbia University College
   of Physicians and Surgeons
New York, New York, U.S.A.
Chapter 3

Helmut Denk
Institute of Pathology
Medical University of Graz
Graz, Austria
Chapter 9

Andrea Fuchsbichler
Institute of Pathology
Medical University of Graz
Graz, Austria
Chapter 9

José L. Jorcano
Department of Molecular
   and Cell Biology
CIEMAT
Madrid, Spain
Email: llanos.casanova@ciemat.es
Chapter 8

Sashi Kesavapany
Laboratory of Neurochemistry,
National Institute of Neurological
   Disorders and Stroke
National Institute of Health
Bethesda, Maryland, U.S.A.
Chapter 4

E. Birgitte Lane
Cancer Research UK Cell Structure
   Research Group
Cell and Developmental Biology Division
University of Dundee
   School of Life Sciences
Dundee, U.K.
Email: e.b.lane@dundee.ac.uk
Chapter 5

M. Fernanda Lara
Department of Molecular
   and Cell Biology
CIEMAT
Madrid, Spain
Chapter 10

Ronald K.H. Liem
Departments of Pathology and Anatomy
    and Cell Biology
Columbia University College
    of Physicians and Surgeons
New York, New York, U.S.A.
Email: rkl2@columbia.edu
Chapter 3

Manuel Navarro
Department of Molecular
    and Cell Biology
CIEMAT
Madrid, Spain
Email: manuel.navarro@ciemat.es
Chapter 6

Harish C. Pant
Laboratory of Neurochemistry
National Institute of Neurological
    Disorders and Stroke
National Institute of Health
Bethesda, Maryland, U.S.A.
Email: panth@ninds.nih.gov
Chapter 4

Denise Paulin
Biologie Moléculaire de la Différenciation
Paris, France
Email: paulin@ccr.jussieu.fr
Chapter 1

Milos Pekny
Department of Medical Biochemistry
Sahlgrenska Academy
    at Göteborg University
Göteborg, Sweden
Email: Milos.Pekny@medkem.gu.se
Chapter 2

Richard H. Quarles
Laboratory of Molecular
    and Cellular Neurobiology
National Institute of Neurological
    Disorders and Stroke
National Institute of Health
Bethesda, Maryland, U.S.A.
Chapter 4

Sergio Ruiz
Cancer Research Center
University of Salamanca
Salamanca, Spain
Chapter 10

Mirentxu Santos
Department of Molecular
    and Cell Biology
CIEMAT
Madrid, Spain
Chapter 10

Carmen Segrelles
Department of Molecular
    and Cell Biology
CIEMAT
Madrid, Spain
Chapter 10

Conny Stumptner
Institute of Pathology
Medical University of Graz
Graz, Austria
Chapter 9

Ulrika Wilhelmsson
Department of Medical Biochemistry
Sahlgrenska Academy
    at Göteborg University
Göteborg, Sweden
Chapter 2

Zhigang Xue
Biologie Moléculaire de la Différenciation
Paris, France
Chapter 1

Kurt Zatloukal
Institute of Pathology
Medical University of Graz
Graz, Austria
Email: kurt.zatloukal@meduni-graz.at
Chapter 9

PREFACE

Molecular biologists' concept of cells is a "biological container" where important molecules float and interact with one another more or less randomly. However, a close view of this "bag" reveals an astonishing net of fibrous proteins crossing throughout the cytoplasm: the cytoskeleton. This structure is composed of three different elements: actin microfilaments, tubulin microtubules, and a third class denoted, due to their size, intermediate filaments (IF). This last network is built by different proteins in different cells types, forming the largest family of cytoskeletal proteins. They are divided into six categories and, except for type V, the lamins, they all form a cytoplasmic web. Types I and II include the epithelial keratins and comprise more than 20 different polypeptides. Type III includes vimentin (expressed in cells of mesenchymal origin), desmin (characteristic of muscle cells), GFAP (in glial cells) and peripherin (in the peripheral nervous system). Type IV IF proteins are found in neurones and include the neurofilament proteins (NF-L, NF-M and NF-H) and α-internexin. Type VI may include, depending on different group criteria, nestin, synemin, paranemin and tanabin. The function of these proteins has long been associated with a structural role. However, this common function does not explain their tissue- and differentiation-specific expression patterns. Evidence is now emerging that IF act as an important framework for the modulation and control of essential cell processes.

In this book, research groups summarize their findings in the IF field in particular focusing on the possible functional roles of IF proteins in cells and their relevance in pathological situations. Paulin's and Pekny's groups summarize these aspects in desmin and GFAP, respectively. In other words, how the functions of muscle cells and astrocytes are dependent on IF. The neurofilaments are covered in two chapters by Pant's and Liem's groups. They focus on mutation and phosphorylation of these proteins and their relationship with neurodegenerative disorders in mouse models and humans.

Keratins, the largest subfamily of IF proteins, are expressed in epithelial cells. Remarkably, the expression of different keratins is strictly regulated in tissue- and differentiation-specific patterns. In this case, data obtained in transgenic mice and genetic analyses of human hereditary syndromes in the early nineties clearly demonstrated that keratins provide cells with mechanical resilience against physical stress. This is reviewed in Lane's chapter. However, there are many intriguing questions to be solved with respect to these proteins. The diversity of keratins is highlighted in the chapter by M. Navarro. He concentrates on the keratin K6 minifamily, in which minimal differences can be observed, discussing the hints about their possible different biological functions. The transcriptional mechanisms regulating keratin expression, which ultimately gives rise to their characteristic expression pattern, are reviewed in M. Blumenberg's chapter.

As commented above, the function of keratins as essential mediators of structural integrity was proposed as a result of the discovery of point mutations in human keratin genes in patients suffering from different epithelial disorders. However, this is only the overall function of the stratified epithelia keratins. Aspects relative to the functions of simple epithelia keratins are covered by two chapters. Zatloukal et al summarize the role of these keratins in chronic liver diseases. Casanova et al review the functionality of keratins K8 and K18 as putative modulators of signaling and apoptosis and their relationship with tumor development and progression. Finally, in the chapter by Santos et al we summarize our findings which suggest that keratin K10 is a mediator of keratinocyte homeostasis.

Overall this book reviews most recent developments in this growing and exciting field and will help those interested in the study of these interesting proteins.

Jesús M. Paramio, Ph.D.

CHAPTER 1

Desmin and Other Intermediate Filaments in Normal and Diseased Muscle

Denise Paulin* and Zhigang Xue

Abstract

The intermediate filament proteins (desmin, vimentin, nestin, synemins and paranemin) synthesized by muscle cells depends on the type of muscle and its stage of development. Desmin is present in all muscles at all stages of development. The others appear transiently or in only certain muscles. The muscles of mice lacking desmin and those of human having a mutated desmin gene that encodes a nonfunctional desmin are abnormal. The severity of the human disease depends on the location of the mutation; it may cause skeletal myopathies, cardiomyopathies or altered vascular elasticity. This report summarizes the function of the desmin gene in the skeletal and smooth muscles, the gene regulation and desmin-related myopathies.

Desmin in Mature Skeletal and Heart Muscle Cells

The cytoskeleton of muscle cells includes proteins whose primary function is to link and anchor structural cell components, especially the myofibrils, the mitochondria, the sarcotubular system and the nuclei.[1] The cytoskeleton has three major filamentous components, intermediate filaments (IFs); microfilaments (actin); and microtubules. The IFs are so named because their diameter (8-10 nm) is intermediate between those of the thick (myosin, 15 nm) and thin (actin, 6 nm) filaments. The cytoskeleton may also be subdivided into the extra-sarcomeric, the intra-sarcomeric and the subsarcolemmal cytoskeleton; the IFs form the extra-sarcomeric cytoskeleton.

Desmin is the main IF protein in mature skeletal and heart muscles. It is encoded by a fully characterized single copy gene,[2] which has been mapped to band q35 to the long arm of human chromosome 2[3] and to band C3 of the mouse to chromosome 1.[4] Desmin is one of the first muscle-specific proteins to be detected in the mammalian embryo, appearing before titin, skeletal muscle actin, myosin heavy chains and nebulin. It can be detected at 8.5 d.p.c. in the developing mouse embryo, in the ectoderm where it is transiently coexpressed with keratin and vimentin.[5] The protein is also found in the heart rudiment on 8.5 d.p.c., and its concentration increases over time, continuing in the myocardial cells during later cardiogenesis.[5] From 9 d.p.c. onwards, desmin can be detected in the myotomes.[6]

The concentration of desmin in skeletal and cardiac muscles remains high throughout embryogenesis and early postnatal life. There is much more desmin in heart muscle cells (2% of total protein) than in skeletal muscle cells (0.35%) of mammal.[7] It forms a three-dimensional scaffold around the myofibrillar Z-disc and interconnects the entire contractile apparatus with the subsarcolemmal cytoskeleton, the nuclei and other cytoplasmic organelles (Fig. 1). Desmin also

*Corresponding Author: Denise Paulin—Biologie Moléculaire de la Différenciation, Paris Cedex, France. Email: paulin@ccr.jussieu.fr

Intermediate Filaments, edited by Jesus Paramio. ©2006 Landes Bioscience and Springer Science+Business Media.

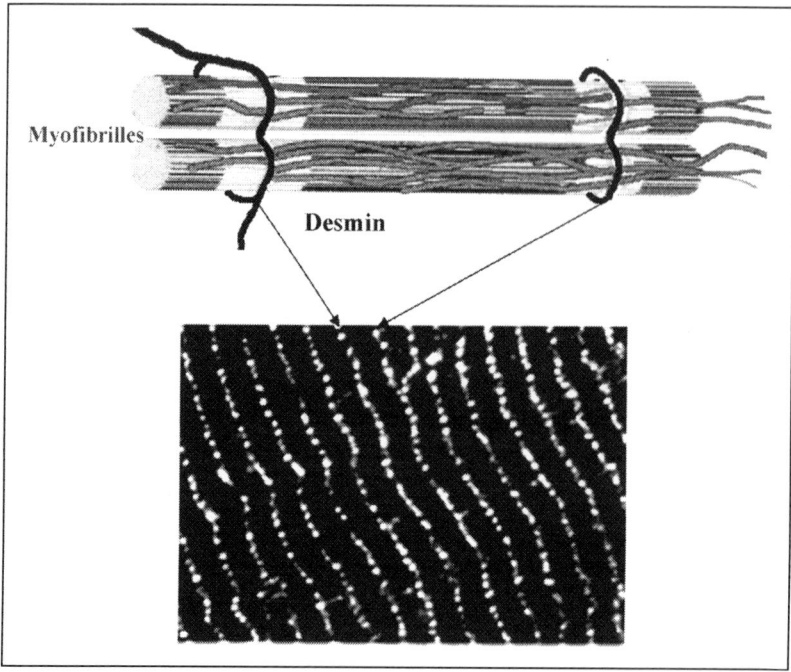

Figure1. Localization of desmin intermediate filaments around the Z-discs in striated muscle. Section was stained with anti-desmin antibody and visualised by immunofluorescence.

forms longitudinal connections between the peripheries of successive Z-disc and along the plasma membrane in chicken and rabbit muscles.[8,9] Desmin is concentrated at the myotendinous junctions and at neuromuscular junctions, in skeletal muscles[10-12] and lies deep within the junctional folds in frog myotendinous junctions, but not immediately subjacent to the junctional membrane.[12] However, desmin is particularly concentrated among and around the ends of the folds in rat neuromuscular junctions.[11] This location is in agreement with the observations of intermediate size filaments between the subneural nuclei and in the postsynaptic folds of freeze-etched frog, rat and snake neuromuscular junctions. Desmin is particularly plentiful in the Purkinje fibres of the heart; the cytoplasm of bovine Purkinje fibres contains 50-75% IFs.[13] Normal cardiomyocytes are also rich in desmin, where it forms a double band structure at intercalated discs, the cell-to-cell contact into which both the longitudinal and transverse IFs are inserted.[14,15] Similarly, desmin seems to be especially abundant at regular intervals along the sarcolemma of both heart and skeletal muscle cells.[14] The membrane proteins vinculin and spectrin are also abundant at the intercalated discs of the Purkinje fibres, in the normal myocardium, and at intervals along the sarcolemma.[16] These two proteins are concentrated in distinct domains at the sarcolemma of skeletal muscles.[17] This location is thought to correspond to the costameres. Thus vinculin and spectrin may act as a link between intracellular structures and the extra cellular matrix.[18] Just how the desmin filaments are anchored to the sarcolemma is not yet known. However, in vitro experiments have shown interactions between desmin and the membrane protein ankyrin.[19]

Desmin and Other IF Proteins in Different Muscles

The profiles of IF proteins in the cells of skeletal, cardiac and smooth muscles vary with the development stage (Table 1). The vimentin gene encodes a 54-kDa cytoskeletal protein. This protein assembles into filaments to form a cytoskeletal network which is attached to the nucleus

Table 1. The intermediate filament proteins

IF Class	Intermediate Filament Protein	Main Tissues	MW	Chromosome Location
Type III	Desmin	Skeletal, heart, smooth	52	2q35(H), 1C3(M)
	Vimentin	Cells of mesenchymal origin	54	10p13(H), 2A2(M)
Type IV	Syncoilin	Skeletal and heart muscles	53.6	1p34.3-p33(H), 4D2.2(M), 5q36(R)
Type VI	Nestin*	Neuroepithelial stem cells, muscle cells		
	Synemin**	Skeletal, heart and smooth muscles	180-150-41	15q26.3(H), 7B5(M)

* also in the lens and CNS. ** also in the lens, CNS, PNS and hematopoietic cells. Chromosome location: H, human; M, mouse; R, rat.

and radiates through the cytoplasm to the plasma membrane. Vimentin is a characteristic IF protein of mesenchymal cells, some of which are the precursors of muscle.[20,21-23] These cells produce desmin when they become committed to a muscle lineage. The 54 kDa protein is the constitutive subunit of the intermediate filaments. It becomes incorporated into the existing vimentin filaments during early myogenesis and forms longitudinal strands. These strands are transformed into transversely organized filaments as the myotubes mature; the filaments lie between the myofibrils at the level of the Z-discs.[22,23] The synthesis of vimentin ceases later in myofibril maturation, and it is not present in normal mature skeletal muscle fibers.[20,21,24]

The IF protein nestin was originally identified in neuroepithelial cells, but is also present in developing skeletal and heart muscles, where it is transiently found with desmin and vimentin.[25,26] Nestin is present in the adult human heart, but its location has not yet been investigated in detail.

Synemin is a large IF protein (150-230 kDa) that is found with desmin and vimentin in muscles and in a few nonmuscle cells. It was originally identified in avian erythrocytes.[27] Synemin was found in porcine muscle,[28] chicken gizzard,[29] chicken and human lens cells.[30,31] It was also observed in a subpopulation of rat astrocytes expressing glial fibrillary acidic protein (GFAP), vimentin and nestin[32] and in some neuroglia together with GFAP, vimentin and keratins.[33] The wide range of species in which it is found suggests that synemin, like nestin, has evolved more rapidly than other IF subunits. This rapid evolution may reflect adaptation. We recently studied the expression of the human synemin gene and identified two major mRNAs coding for the two IF isoforms (180 and 150 kDa) that are found in the human skeletal and smooth muscles.[34] The M isoform (also called desmuslin by ref. 35) lacks the 312 aminoacids encoded by intron IV.

We analyzed the spatial and temporal expression of synemin gene during mouse development to obtain a clearer picture of how the synthesis of the synemin isoforms is regulated. All, three isoforms of mouse synemin were isolated. We focused on the formation of the synemin IF network in tissues from mice lacking desmin and vimentin because synemin appears to be unable to assemble alone; it requires other IF partners to form filaments. Our experiment showed that synemin does not form filaments in the skeletal or cardiac muscles in the absence of desmin. The situation in smooth muscles where vimentin and desmin are present in the same cell is more complex. The absence of desmin from bladder or blood vessels does not influence the distribution of synemin, whereas the absence of vimentin disrupts synemin organization. This could reflect differences in the network-forming abilities of desmin and vimentin in smooth muscle and in their relationship with their partners.[36]

Table 2. The intermediate filament protein functions

Desmin	In skeletal muscle between the peripheries of successive Z-discs and along the plasma membrane Myotendinous junction Neuromuscular junction
Syncoilin	Neuromuscular junction Sarcolemma Z-discs

Nestin	Neuromuscular junction:
	  Between the post-junctional folds and the subneural nuclei
	  Between the nucleus and the myofibrillar cytoskeleton
	Myotendinous junction:
	  Between myofibrils at the Z-disc level
	  Longitudinal strands close to and at the junction

Synemin	With desmin in Z-disc associated striations and at the sarcolemma Also at the myotendinous junction and in the post-synaptic area at the motor end plate

Functions of Desmin in Striated Muscles (Table 2)

A null mutation was introduced into the mouse desmin gene by homologous recombination. The resulting desmin knockout mice (Des-/-) develop normally and are fertile. However, defects develop in the skeletal, smooth and cardiac muscles of the newborn mice.[37,38] We have carried out a detailed analysis of somitogenesis, muscle formation, maturation, degeneration, and regeneration in Des-/- mice. The results demonstrate that the muscle differentiation and cell fusion at all the early stages occur normally. However, modifications begin to develop after birth, essentially in the weight-bearing muscles, such as the soleus, and in continually used muscles like the diaphragm and the heart.[39] The mice lacking desmin are weaker and fatigue more easily than their normal littermates. The absence of desmin renders these fibers more susceptible to damage during contraction. The myofibers degenerate, and this is followed by macrophage infiltration, and then by regeneration. These cycles of degeneration and regeneration result in an increase in the relative proportion of slow myosin heavy chain (MHC) and a decrease in fast MHC. This second wave of myofibrillogenesis during regeneration was often aberrant and showed signs of disorganization. Mitochondria also accumulated below the sarcolemma in these muscles. The lack of desmin was not offset by an increase in vimentin in these mice, during either development or regeneration. The absence of desmin filaments from the sarcomeres did not interfere with primary muscle formation or regeneration. However, myofibrillogenesis in regenerating fibers was often abortive, indicating that desmin is implicated in this repair process. Thus, desmin is essential for maintaining the structural integrity of highly solicited skeletal muscle, but not for myogenic commitment, differentiation, or the fusion of skeletal muscle.

The hearts of mice lacking desmin are myopathic, with impaired active force generation, although the wall compliance was not altered. Diastolic pressure was increased at all filling volumes. Since passive wall stress was unchanged, the alteration in distolic pressure was due to the thicker ventricle wall. Hence, desmin filaments may have a role in the generation of active force by cardiac muscle, possibly by supporting sarcomere alignment or force transmission.[40]

The mitochondria in the hearts of mice lacking desmin are abnormally shaped and distributed. The conventional kinesin, the microtubule-associated plus end-directed motor was not associated with the mitochondria in desmin null hearts, suggesting that the positioning of mitochondria is dynamic, involving the desmin filaments, the motor kinesin and the microtubule network. The hearts of mice lacking desmin also have above-normal creatine kinase activity, below-normal cytochrome c and redistributed Bcl2.[41]

Function of Desmin in Smooth Muscles

The intermediate filament system is a major constituent of the smooth muscle cytoskeleton. The intermediate filaments in smooth muscle are composed mainly of desmin, vimentin, and synemin, but other intermediate filament proteins, like the cytokeratins, have also been found.[42] Smooth muscles can contain both vimentin and desmin simultaneously.[43] Nevertheless, as there appears to be no difference in their functions, it is not clear why certain smooth muscles contain mainly desmin intermediate filaments, while others have mainly vimentin. A lack of desmin leads to a decrease in tension development in the urinary bladder and vas deferens and a loss of shortening velocity.[44]

The intermediate filaments of smooth muscle are associated with dense plaques in the membrane, and dense bodies in the cytoplasm,[45] which are considered to be the smooth muscle equivalents of the Z-discs.

The desmin intermediate filaments bind to dense bodies in smooth muscle, and might help anchor the contractile units. The visceral smooth muscles of desmin-deficient mice develop significantly less active force, suggesting that the desmin intermediate filaments help transmit active force.[44]

Desmin and the Passive and Active Tension of Blood Vessels

Desmin is involved in smooth muscle dilation and contraction in resistance arteries, but not in large arteries.[46] Larger arteries contain few desmin-positive cells, whereas muscle arteries and veins have many desmin-positive cells. This gradient in the vimentin/desmin ratio in the vascular tree also occurs in human vascular tissues.[47] Vimentin is typically found in mesenchymal nonmuscle cells; it is also the major intermediate filament protein in large arteries.[48] The desmin /vimentin ratio varies with the size of arteries; large arteries have relatively large amounts of vimentin and small amounts of desmin, while the opposite is true in smaller arteries.[47,48]

Arteries with a diameter of 100-300 μm play a key role in determining the resistance to blood flow[49] and their physiological function is probably influenced by their passive and active mechanical properties. If the desmin/vimentin concentration gradient that occurs in larger arterial vessels also extends to resistance arteries, it is likely that they contain many desmin intermediate filaments and they may be most important in the smaller arteries.

The main features of the large arteries of Des-/- mice are disorganization of SMC, cell hypoplasia and a loss of adhesion, mainly in the thoracic aorta, which is particularly exposed to mechanical stress and affected by fatigue. The carotid arteries of these mice are more viscous and less distensible than those of Des+/+.[50]

Mice lacking desmin suffer from microvascular dysfunction. The resulting smooth muscle hyporeactivity shows a link between the absence of desmin and exaggerated structural adaptation (remodeling) in response to changes in blood flow. This might be due unbalanced flow-dependent remodeling or to the higher blood flow needed by tissues supplied by resistance arteries.[46] The microvascular defects that occur in the absence of desmin may be important cause of the functional damage ocurring in desmin-related myopathies in humans.

Gene Regulation

The expression of the muscle-specific intermediate filament desmin gene in skeletal muscle is controlled in part by a muscle-specific enhancer. This activity can be divided into myoblast-specific and myotube-specific activation domains. The myotube-specific region contains MyoD and MEF2 sites, whereas the myoblast-specific region contains Sp1, Krox and Mb sites. A novel site located between the MyoD and MEF2 sites (named Mt) is required for full transcription activity (Fig. 2).

A new type of combined CArG/octamer plays a prominent role in the regulation of the desmin gene in arterial smooth muscle cells. This element can bind to the serum response factor (SRF) and an Oct-like factor.[51]

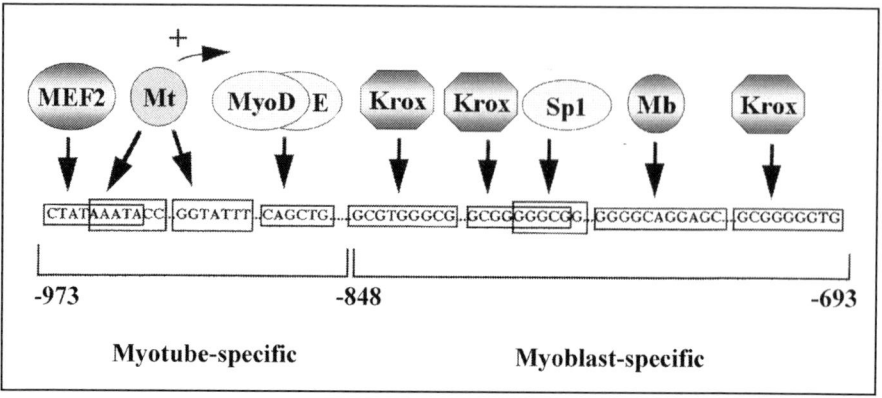

Figure 2. Illustration of specific enhancer domains of skeleton muscle in the upstream of desmin gene. The myotube-specific enhancer domain contains MEF2 and heterodimere type factors (MyoD or Myf5; E12 or E47). The factor Mt is required for full transcription activity. The myoblast-specific enhancer domain binds SP1, Krox type factors and an unknown Mb factor.

The IF protein vimentin is replaced by desmin when myogenesis occurs in vitro; it is the result of a switch in gene transcription. The vimentin promoter is recognized and activated by a protein that is probably identical to HTAF-1. This factor is present in proliferating myoblasts, but disappears when these cells fuse to form multinucleated myotubes. HTAF-1 is a differentiation stage-specific factor that turns off the expression of the vimentin gene during myogenesis.[25]

Desmin-Related Myopathies (DRM)

Desmin-related myopathies (DRM) are caused by defects in the formation of the desmin filament network; they include clinically diverse sporadic and familial muscle diseases involving the degeneration of skeletal and cardiac fibers[52-54] (Fig. 3). The DRMs have a wide genetic spectrum, with mutations that have been mapped to various parts of the human desmin gene. A total of 21 mutations or deletions in the desmin gene have been reported to date, including 16 misense mutations, three small in-frame deletion of 1-7 amino acids and the insertion of a single nucleotide resulting in translation termination, one mutation in splice donor or acceptor sites flanking exon 3 resulting in deletion of 32 amino acids.[54] A missense mutation in the alpha-B crystallin gene, which encodes the chaperone protein, disrupts the organization of desmin filaments causing them to aggregate and give rise to clinical symptoms similar to those of DRMs, including distal myopathy and cardiomyopathy[55,56] (Fig. 4). Two other IF proteins, synemin and syncoilin, may also be involved in DRM, although no mutation has yet been found in the genes encoding these two proteins. It has been suggested that synemin is a component of heteropolymeric IF and may be important for cytoskeletal cross-linking. The muscles contain synemin and paranemin, together with plectin, in the same location as desmin in the striations associated with the Z-disc and at the sarcolemma, where they display topological and structural relationships. Furthermore, the desmin cytoskeleton is disorganized in plectin-related muscular dystrophy.

In conclusion, the expression of IF genes during development and differentiation, the characterization of IF protein network and the knock-out mice can help us to reveal the fundamental role of desmin filaments in cell architecture, myofibrillar organization and physiological function in muscle.

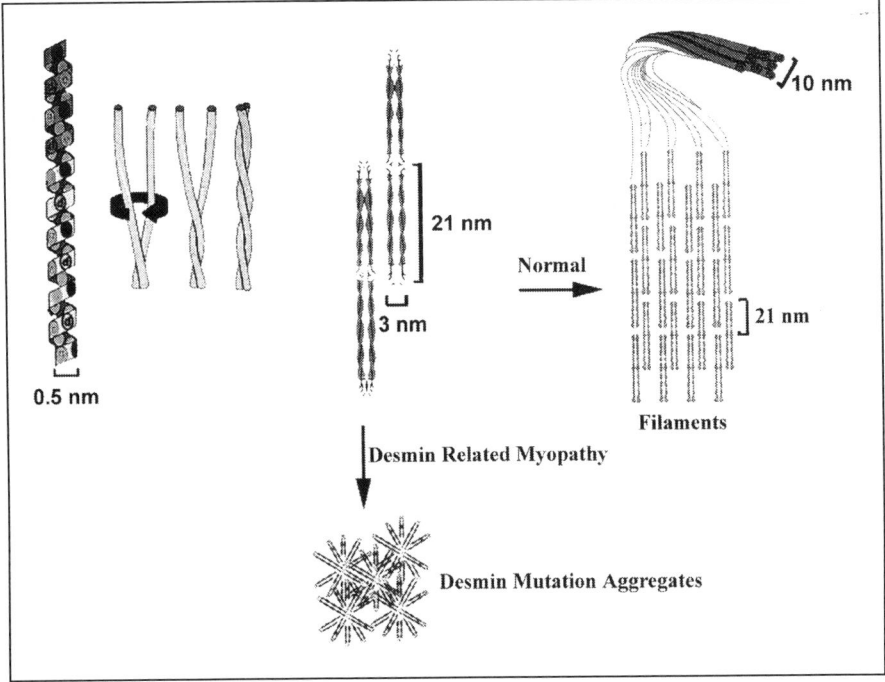

Figure 3. The normal filament polymerization and the desminopathy aggregate formation in striated muscle cells.

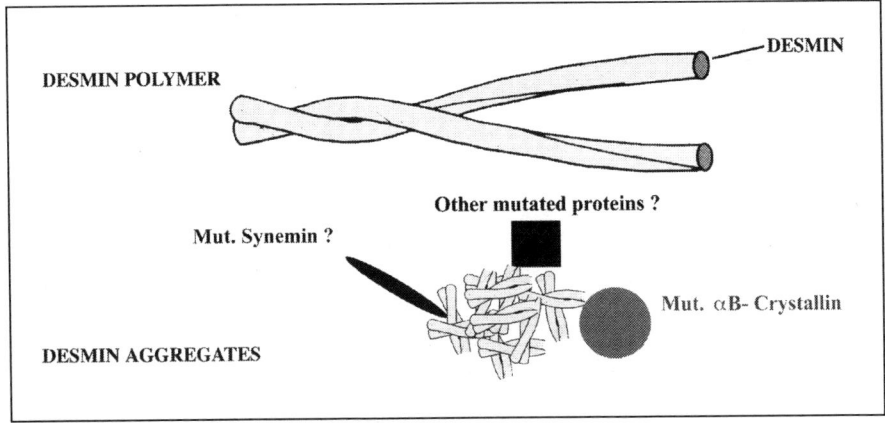

Figure 4. Desmin-related myopathies causing by mutations in associated proteins or in other intermediate filaments.

References

1. Stromer MH. The cytoskeleton in skeletal, cardiac and smooth muscle cells. Histol Histopathol 1998; 13(1):283-291.
2. Li ZL, Lilienbaum A, Butler-Browne G et al. Human desmin-coding gene: Complete nucleotide sequence, characterization and regulation of expression during myogenesis and development. Gene 1989; 78(2):243-254.
3. Viegas-Pequignot E, Li ZL, Dutrillaux B et al. Assignment of human desmin gene to band 2q35 by nonradioactive in situ hybridization. Hum Genet 1989; 83(1):33-36.
4. Li ZL, Mattei MG, Mattei JF et al. Assignment of the mouse desmin gene to chromosome 1 band C3. Genet Res 1990; 55(2):101-105.
5. Schaart G, Viebahn C, Langmann W et al. Desmin and titin expression in early postimplantation mouse embryos. Development 1989; 107(3):585-596.
6. Mayo ML, Bringas Jr P, Santos V et al. Desmin expression during early mouse tongue morphogenesis. Int J Dev Biol 1992; 36(2):255-263.
7. Price MG. Molecular analysis of intermediate filament cytoskeleton—a putative load-bearing structure. Am J Physiol 1984; 246(4 Pt 2):H566-572.
8. Lazarides E, Hubbard BD. Immunological characterization of the subunit of the 100 A filaments from muscle cells. Proc Natl Acad Sci USA 1976; 73(12):4344-4348.
9. Wang K, Ramirez-Mitchell R. A network of transverse and longitudinal intermediate filaments is associated with sarcomeres of adult vertebrate skeletal muscle. J Cell Biol 1983; 96(2):562-570.
10. Askanas V, Bornemann A, Engel WK. Immunocytochemical localization of desmin at human neuromuscular junctions. Neurology 1990; 40(6):949-953.
11. Sealock R, Murnane AA, Paulin D et al. Immunochemical identification of desmin in Torpedo postsynaptic membranes and at the rat neuromuscular junction. Synapse 1989; 3(4):315-324.
12. Tidball JG. Desmin at myotendinous junctions. Exp Cell Res 1992; 199(2):206-212.
13. Thornell LE, Eriksson A. Filament systems in the Purkinje fibers of the heart. Am J Physiol 1981; 241(3):H291-305.
14. Tokuyasu KT. Visualization of longitudinally-oriented intermediate filaments in frozen sections of chicken cardiac muscle by a new staining method. J Cell Biol 1983; 97(2):562-565.
15. Thornell LE, Butler-Browne GS, Carlsson E et al. Cryoultramicrotomy and immunocytochemistry in the analysis of muscle fine structure. Scan Electron Microsc 1986; (Pt 4):1407-1418.
16. Thornell LE, Johansson B, Eriksson A et al. Intermediate filament and associated proteins in the human heart: An immunofluorescence study of normal and pathological hearts. Eur Heart J 1984; 5(Suppl F):231-241.
17. Porter GA, Dmytrenko GM, Winkelmann JC et al. Dystrophin colocalizes with beta-spectrin in distinct subsarcolemmal domains in mammalian skeletal muscle. J Cell Biol 1992; 117(5):997-1005.
18. Shear CR, Bloch RJ. Vinculin in subsarcolemmal densities in chicken skeletal muscle: Localization and relationship to intracellular and extracellular structures. J Cell Biol 1985; 101(1):240-256.
19. Georgatos SD, Weber K, Geisler N et al. Binding of two desmin derivatives to the plasma membrane and the nuclear envelope of avian erythrocytes: Evidence for a conserved site-specificity in intermediate filament-membrane interactions. Proc Natl Acad Sci USA 1987; 84(19):6780-6784.
20. Barbet JP, Thornell LE, Butler-Browne GS. Immunocytochemical characterisation of two generations of fibers during the development of the human quadriceps muscle. Mech Dev 1991; 35(1):3-11.
21. Furst DO, Osborn M, Weber K. Myogenesis in the mouse embryo: Differential onset of expression of myogenic proteins and the involvement of titin in myofibril assembly. J Cell Biol 1989; 109(2):517-527.
22. Gard DL, Lazarides E. The synthesis and distribution of desmin and vimentin during myogenesis in vitro. Cell 1980; 19(1):263-275.
23. Tokuyasu KT, Maher PA, Dutton AH et al. Intermediate filaments in skeletal and cardiac muscle tissue in embryonic and adult chicken. Ann NY Acad Sci 1985; 455:200-212.
24. Moura-Neto V, Kryszke MH, Li Z et al. A 28-bp negative element with multiple factor-binding activity controls expression of the vimentin-encoding gene. Gene 1996; 168(2):261-266.
25. Kachinsky AM, Dominov JA, Miller JB. Myogenesis and the intermediate filament protein, nestin. Dev Biol 1994; 165(1):216-228.
26. Sjoberg G, Jiang WQ, Ringertz NR et al. Colocalization of nestin and vimentin/desmin in skeletal muscle cells demonstrated by three-dimensional fluorescence digital imaging microscopy. Exp Cell Res 1994; 214(2):447-458.
27. Granger BL, Repasky EA, Lazarides E. Synemin and vimentin are components of intermediate filaments in avian erythrocytes. J Cell Biol 1982; 92(2):299-312.
28. Bilak SR, Sernett SW, Bilak MM et al. Properties of the novel intermediate filament protein synemin and its identification in mammalian muscle. Arch Biochem Biophys 1998; 355(1):63-76.

29. Bellin RM, Sernett SW, Becker B et al. Molecular characteristics and interactions of the intermediate filament protein synemin. Interactions with alpha-actinin may anchor synemin-containing heterofilaments. J Biol Chem 1999; 274(41):29493-29499.
30. Granger BL, Lazarides E. Expression of the intermediate-filament-associated protein synemin in chicken lens cells. Mol Cell Biol 1984; 4(10):1943-1950.
31. Tawk M, Titeux M, Fallet C et al. Synemin expression in developing normal and pathological human retina and lens. Exp Neurol 2003; 183(2):499-507.
32. Sultana S, Sernett SW, Bellin RM et al. Intermediate filament protein synemin is transiently expressed in a subset of astrocytes during development. Glia 2000; 30(2):143-153.
33. Hirako Y, Yamakawa H, Tsujimura Y et al. Characterization of mammalian synemin, an intermediate filament protein present in all four classes of muscle cells and some neuroglial cells: Colocalization and interaction with type III intermediate filament proteins and keratins. Cell Tissue Res 2003; 313(2):195-207.
34. Titeux M, Brocheriou V, Xue ZG et al. Human synemin gene generates splice variants encoding two distinct intermediate filament proteins. Eur J Biochem 2001; 268(24):6435-6449.
35. Mizuno Y, Thompson TG, Guyon JR et al. Desmuslin, an intermediate filament protein that interacts with alpha -dystrobrevin and desmin. Proc Natl Acad Sci USA 2001; 98(11):6156-6161.
36. Xue ZG, Cheraud Y, Brocheriou V et al. The mouse synemin gene encodes three intermediate filament proteins generated by alternative exon usage and different open reading frames. Exp Cell Res 2004; 298:431-444.
37. Li Z, Colucci-Guyon E, Pincon-Raymond M et al. Cardiovascular lesions and skeletal myopathy in mice lacking desmin. Dev Biol 1996; 175(2):362-366.
38. Milner DJ, Weitzer G, Tran D et al. Disruption of muscle architecture and myocardial degeneration in mice lacking desmin. J Cell Biol 1996; 134(5):1255-1270.
39. Li Z, Mericskay M, Agbulut O et al. Desmin is essential for the tensile strength and integrity of myofibrils but not for myogenic commitment, differentiation, and fusion of skeletal muscle. J Cell Biol 1997; 139(1):129-144.
40. Balogh J, Merisckay M, Li Z et al. Hearts from mice lacking desmin have a myopathy with impaired active force generation and unaltered wall compliance. Cardiovasc Res 2002; 53(2):439-450.
41. Linden M, Li Z, Paulin D et al. Effects of desmin gene knockout on mice heart mitochondria. J Bioenerg Biomembr 2001; 33(4):333-341.
42. Lazarides E. Intermediate filaments: A chemically heterogeneous, developmentally regulated class of proteins. Annu Rev Biochem 1982; 51:219-250.
43. Osborn M, Caselitz J, Puschel K et al. Intermediate filament expression in human vascular smooth muscle and in arteriosclerotic plaques. Virchows Arch A Pathol Anat Histopathol 1987; 411(5):449-458.
44. Sjuve R, Arner A, Li Z et al. Mechanical alterations in smooth muscle from mice lacking desmin. J Muscle Res Cell Motil 1998; 19(4):415-429.
45. Kargacin GJ, Cooke PH, Abramson SB et al. Periodic organization of the contractile apparatus in smooth muscle revealed by the motion of dense bodies in single cells. J Cell Biol 1989; 108(4):1465-1475.
46. Loufrani L, Matrougui K, Li Z et al. Selective microvascular dysfunction in mice lacking the gene encoding for desmin. Mechanical properties and structure of carotid arteries in mice lacking desmin. FASEB J 2002; 16(1):117-119.
47. Johansson B, Eriksson A, Virtanen I et al. Intermediate filament proteins in adult human arteries. Anat Rec 1997; 247(4):439-448.
48. Frank ED, Warren L. Aortic smooth muscle cells contain vimentin instead of desmin. Proc Natl Acad Sci USA 1981; 78(5):3020-3024.
49. Christensen KL, Mulvany MJ. Location of resistance arteries. J Vasc Res 2001; 38(1):1-12.
50. Lacolley P, Challande P, Boumaza S et al. Mechanical properties and structure of carotid arteries in mice lacking desmin. Cardiovasc Res 2001; 51(1):178-187.
51. Mericskay M, Parlakian A, Porteu A et al. An overlapping CArG/octamer element is required for regulation of desmin gene transcription in arterial smooth muscle cells. Dev Biol 2000; 226(2):192-208.
52. Goebel HH, Fardeau M. Desmin - protein surplus myopathies. Neuromuscul Disord 2002; 12(7-8):687-692.
53. Goldfarb LG, Vicart P, Goebel HH et al. Desmin myopathy. Brain 2004; 127(Pt 4):723-734.
54. Paulin D, Huet A, Khanamyrian L et al. Desminopathies in muscle disease. J Pathol Annu Rev 2004; 204:418-427.
55. Vicart P, Caron A, Guicheney P et al. A missense mutation in the alphaB-crystallin chaperone gene causes a desmin-related myopathy. Nat Genet 1998; 20(1):92-95.
56. Selcen D, Engel AG. Myofibrillar myopathy caused by novel dominant negative alpha B-crystallin mutations. Ann Neurol 2003; 54(6):804-810.

CHAPTER 2

Intermediate Filaments in Astrocytes in Health and Disease

Milos Pekny* and Ulrika Wilhelmsson

Abstract

Astroglial cells (Fig. 1) are the most abundant cells in the mammalian central nervous system (CNS), and we only now start to fully realize their importance both in health and disease. Upregulation of intermediate filaments (IFs) has been a well-known hallmark of astrocytes activated by a disease process. This chapter focuses on the function of IF proteins and IFs in astroglial cells both in health and in pathological situations, such as severe mechanical or osmotic stress, hypoxia, brain and spinal cord injury as well as in CNS regeneration.

IFs in Astroglial Cells during Development

The first astroglial cells to appear during development of the CNS are the radial glia, proliferating precursor cells giving birth to neurons and glial cells. The radial processes of these cells span the entire thickness of the neural tube and function as migratory paths for the immature neurons (reviewed in ref. 1). In lower vertebrates the radial glia persist into adulthood,[2] while in mammals most of them are transformed into astrocytes after the morphogenesis process is accomplished around birth.[3,4] In some places though the radial glia persist into adulthood; in the cerebellum as Bergmann glia[5] and in the retina as Müller cells[6] (Fig. 2). In the early development, the IFs in the radial glia are composed of nestin[7] and vimentin.[8,9] In primates, the IF network in radial glia also contain glial fibrillary acidic protein (GFAP) primary localized in the cellular main processes.[10]

Around birth, the radial glia of the mammalian CNS transform into astrocytes and the expression of vimentin decreases while the expression of GFAP gradually increases.[11] In mice, the amount of GFAP mRNA in the brain increases postnatally and reaches the peak at postnatal day 15 (P15) when cell proliferation declines.[12,13] The IF protein synemin has recently been reported to be transiently expressed in some immature astrocytes.[14] The levels of nestin do not seem to be altered until the transition from radial glia into adult astrocytes is completed and then it is down-regulated.[15] The IF network of mature astrocytes is thereby composed of GFAP as the major IF protein and vimentin ranging from very low to intermediate levels depending on the subpopulation of astrocytes.[16,17]

Mature astrocytes have fine processes extending from the main cellular processes and they give the cell a characteristic bushy appearance (Fig. 1a). The IF network however, is restricted to the main processes and the soma of astrocytes[18,19] (Fig. 3).

The most recent reports show that GFAP-positive astroglial cells might be involved in the baseline neurogenesis in the adult mammalian CNS. Gage and coworkers suggested that astrocytes positively control neurogenesis in the two regions of the adult CNS in which new neurons

*Corresponding Author: Milos Pekny—Department of Medical Biochemistry, Sahlgrenska Academy at Göteborg University, Box 440, 405 30 Göteborg, Sweden.
Email: Milos.Pekny@medkem.gu.se

Intermediate Filaments, edited by Jesus Paramio. ©2006 Landes Bioscience and Springer Science+Business Media.

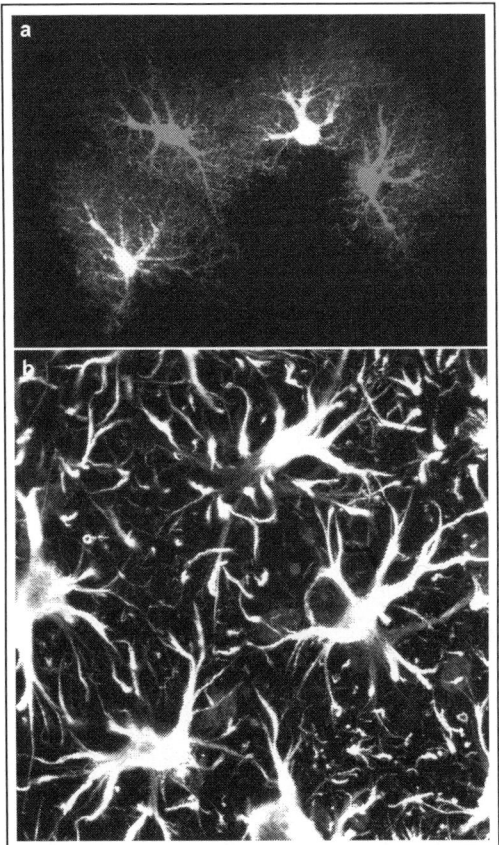

Figure 1. Astrocytes in the adult mammalian CNS. a) astrocytes in the hippocampus filled with two different dyes (Alexa 568 and Lucifer yellow). The CNS is divided into domains and each of them is accessed by fine cellular processes of one individual astrocyte (from ref. 175, the courtesy of Wilhelmsson, Bushong, Ellisman and Pekny). b) astrocytes in the cortex visualized by antibodies against GFAP.

are generated even in the adult, i.e., in the dentate gyrus of the hippocampus and in the subventricular zone.[20] Most interestingly, it was proposed that the majority of neural stem cells in the adult CNS are at some point GFAP positive, i.e., could be defined as astroglial cells.[21-24] Thus, perhaps rather unexpectedly, astroglial cells might be both an important cell type controlling adult neurogenesis as well as the precursors to all neurons that are added in the adult life.

GFAP and Its Regulation

GFAP was identified in 1971 as a protein found in the CNS of patients suffering from multiple sclerosis[25] and during the last 30 years it has been used as the primary marker of mature astrocytes in the CNS.[26] The human *GFAP* gene is located on chromosome 17,[27] while the mouse gene lies on chromosome 11.[28] The GFAP gene is highly conserved among species.[29,30] The transcription of *GFAP* is controlled through several different regulatory elements, which have been described in both humans and rodents.[31-36] About 2 kb of both the human and rodent *GFAP* promoter were shown to be sufficient to direct the expression of reporter genes to astrocytes,[37-39] and constructs containing variable length of the *GFAP* promoter have been used to direct gene expression to astroglial cells as well as to GFAP-positive cells outside the CNS (for review see ref. 40). Several groups have mapped regulatory elements to a distal and a proximal

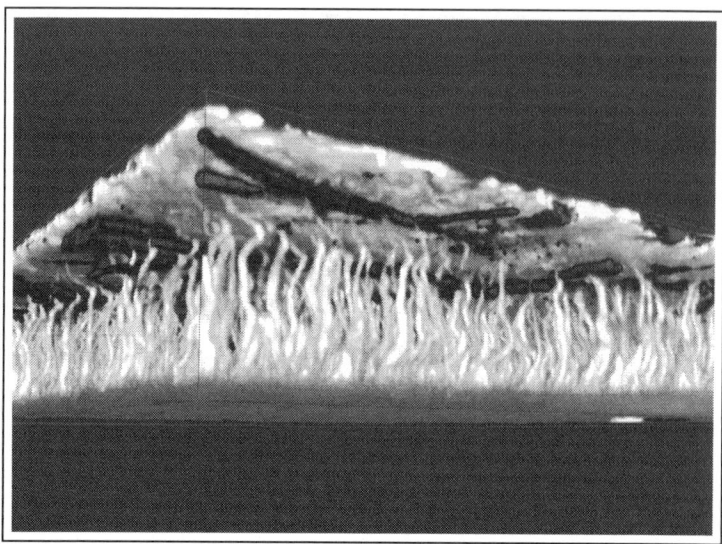

Figure 2. Radial glia in the adult mammalian retina. During development of the CNS, radial glia guide neurons into their final destinations. Later on, most radial glia differentiate into various astroglial cell types, however in the retina and cerebellum, they persist into adulthood as Müller cells and Bergmann glia, respectively. Parallel arrays of Müller cells and a network of astrocytes, both visualized by antibodies against GFAP (white and gray), vessels visualized by isolectin (black). The picture is used with the courtesy of Lundkvist and Pekny.

region within this sequence.[31,34,41] One important site is the consensus AP-1 sequence, a binding site for the Fos and Jun families of transcription factors.[33,41,42] Several methylation sites in the *GFAP* gene were suggested to control cell differentiation. In general, the *GFAP* gene is methylated less in neural than in nonneural tissues,[43-46] and it was proposed that *GFAP* promoter region at a position -1176 is demethylated during rodent development in cells of neuroectodermal origin in the CNS and later remethylated in the mature cells.[45,47] The demethylation and remethylation of a binding element for transcription factors, such as STAT3 might induce differentiation towards a glial lineage.[48] Recently, Song and Ghosh revealed another regulatory mechanism of GFAP expression. They reported that FGF-2 can facilitate the access for transcription

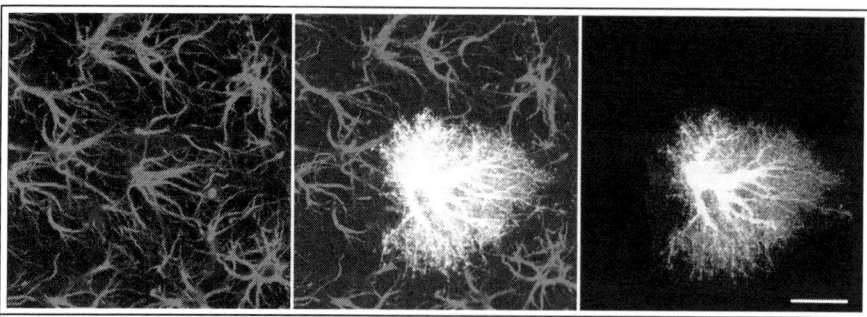

Figure 3. Three-dimensional reconstruction of reactive astrocytes after dye filling (centre and right) shows the typical bushy appearance of astrocytes with fine cellular processes which cannot be visualized by antibodies against GFAP (left and centre). Scale bar, 20 μm. Reproduced from reference 120, ©2004 with permission from the Society for Neuroscience.

factor STAT to the promoter binding site by inducing chromatin remodelling at the *GFAP* promoter through altered methylation of histone H3.[49]

Vimentin and *Nestin* and Their Regulation

Apart from astroglial cells, vimentin is found mainly in cells of mesenchymal origin.[50] Nestin is widely expressed during development in the nervous system and in the developing muscle[51,52] as well as in the adult CNS (e.g., in neural stem cells, reactive astrocytes and endothelial cells[7,53,54]). Several regulatory elements in the *vimentin* promoter region have been described,[55-58] but their involvement in astrocytic expression of vimentin remains incompletely understood. Similar to the *GFAP* gene, the *vimentin* promoter region contains the AP-1 sequence that activates vimentin expression through the Jun and Fos pathways.[59,60] Silencer elements with binding sites for regulatory factors Sp1 and ZBP-89[61-64] as well as an NF kappa B binding site[65] have been identified.

The second intron of the *nestin* gene contains enhancer elements important for nestin expression in the embryonic CNS.[66-70] In the adult CNS, a 636 bp region of the second intron in the rat *nestin* gene is sufficient for nestin expression in the neurogenic zones, but not for a complete response in nestin expression in reactive astrocytes upon injury.[71]

Formation of IFs in Astroglial Cells

GFAP monomers rapidly assemble into di-, tetra-, or larger subunits of polymers that can be incorporated into the IF network. The dynamic feature of the IF network depends both upon the equilibrium between filaments and unassembled subunits and the regulation of filament assembly/disassembly by phosphorylation of the head domain of the IF proteins. The IFs were first considered to be rather static structures primarily responsible for maintaining the cell shape.[72,73] However, later findings both in vitro[74,75] and in vivo[76-79] revealed that the IFs are more dynamic structures and are in a dynamic equilibrium with the pool of soluble subunits (reviewed in ref. 80). Vikström and coworkers assessed the turnover of vimentin subunits in IF fibers in vitro by using the FRAP technique. Rhodamine-labeled vimentin was injected into fibroblasts and was readily incorporated into the endogenous IF network. After bleaching the fluorescent IF fibers with a laser beam, fluorescence returned to the IF fibers throughout their length within a few minutes, proving the existence of a pool of subunits and unpolymerized IF proteins that are in a dynamic equilibrium with the IF network.[78] More recent reports have shown that nonfilamentous IF protein can be rapidly transported along the microtubule tracks,[81,82] implying a complex crosstalk between different cytoskeletal systems.[83]

Phosphorylation of several serine and threonine residues, predominantly on the N-terminal head domain of the GFAP molecule, is important for the rearrangements of the IF network in situations such as cell motility.[84-87] Various kinases, such as cdc2 kinase, protein kinase A, protein kinase C, Ca^{2+}/calmodulin-dependent protein kinase II and Rho kinase, all phosphorylate GFAP and thereby both increase the disassembly of IFs and inhibit the filament assembly.[88-91] These events increase the pool of free, phosphorylated monomers that can readily be reassembled after dephosphorylation by phosphatases. Even though this implicates the phosphorylation as a general mechanism to regulate the equilibrium and turnover rate of different pools of IFs, the distinct phosphorylation sites and kinases may allow more specific actions,[87,92] as for example Rho kinase phosphorylating GFAP in cytokinesis.[93] Different subpopulations of astrocytes in vivo seem to contain different levels of phosphorylated GFAP, suggesting a role for phosphorylation in the nondividing astroglial cells.[94]

IFs in vivo are often, if not always, heteropolymeric. For determining the partnership in the formation of IF heteropolymers in astrocytes, transgenic mice deficient for individual IF proteins were instrumental. In nonreactive astrocytes, GFAP and vimentin form IFs while in reactive astrocytes, nestin can be found as the additional partner in the IF network[95] (Table 1). The studies of astrocytes lacking GFAP and/or vimentin have revealed that GFAP can form IFs on its own in vimentin deficient ($Vim^{-/-}$) astrocytes, but such filaments form more compact bundles than in wild-type astrocytes (Fig. 4, Table 2), suggesting that at least a low level of

Table 1. Composition of IFs in nonreactive and reactive astrocytes of wild-type mice and mice deficient in GFAP and/or vimentin

	Composition of IFs		Reactive Astrocytes:
Genotype	Nonreactive Astrocytes	Reactive Astrocytes	IF Amount/Bundling
Wild-type	GFAP, vimentin	GFAP, vimentin, nestin	Normal/normal
GFAP$^{-/-}$	No IFs (nonfilamentous vimentin)	Vimentin, nestin	Decreased/normal
Vim$^{-/-}$	GFAP	GFAP (nonfilamentous nestin)	Decreased/tight
GFAP$^{-/-}$Vim$^{-/-}$	No IFs	No IFs (nonfilamentous nestin)	–

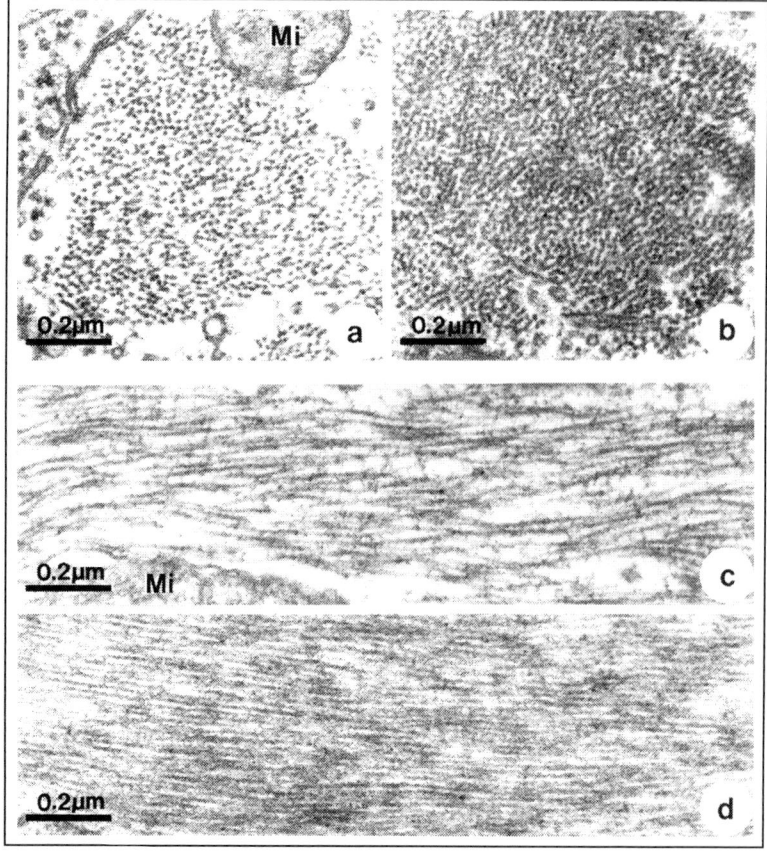

Figure 4. Compared to wild-type astrocytes (a,c), in *Vim*$^{-/-}$ astrocytes (b,d), IFs are composed of GFAP and form more densely packed bundles with the distance between the adjacent IFs being reduced. These electron micrographs show cytoplasmic details from astrocytes in the dorsal funiculus of the cervical spinal cord of healthy adult mice. The IF bundles were sectioned transversally (a,b) and longitudinally (c,d). Mi, mitochondrion. Reproduced, with permission, from reference 96.

Table 2. *Quantitative comparison of the density of IFs within IF bundles in wild-type and Vim$^{-/-}$ astrocytes in the intact CNS*

	Mice	Mean ± SEM	Significance
Number of IFs/0.1 μm² (cross-section)	Wild-type	215 ± 8	
	Vim$^{-/-}$	334 ± 7	$p < 0.0001$
Distance within a bundle between IFs (nm)	Wild-type	11.8 ± 0.5	
	Vim$^{-/-}$	4.9 ± 0.3	$p < 0.0001$

vimentin is needed for normal IF formation in the astrocytes.[96-98] Studies in mice deficient in GFAP (*GFAP$^{-/-}$*) showed that vimentin does not seem to form IF on its own, or it does so only with a very low efficiency[99,100] (Fig. 5). In contrast, the reactive *GFAP$^{-/-}$* astrocytes contain IFs since vimentin can polymerize with nestin, which is expressed in reactive astrocytes.[96] GFAP does not polymerize with nestin in reactive *Vim$^{-/-}$* astrocytes and consequently, the IFs seem to contain only GFAP and they exhibit the characteristic tight bundling similar to *Vim$^{-/-}$* nonreactive astrocytes. In reactive astrocytes lacking both GFAP and vimentin (*GFAP$^{-/-}$Vim$^{-/-}$*) no IFs are formed, and the nestin protein which is produced, stays in a nonfilamentous form[96] (Fig. 6 and Table 1). Nestin was proposed to facilitate phosphorylation-dependent disassembly of vimentin IFs during mitosis and was suggested to play a role in the distribution of IF protein to daughter cells.[101] The IF protein synemin was detected in some astroglial cell populations that express

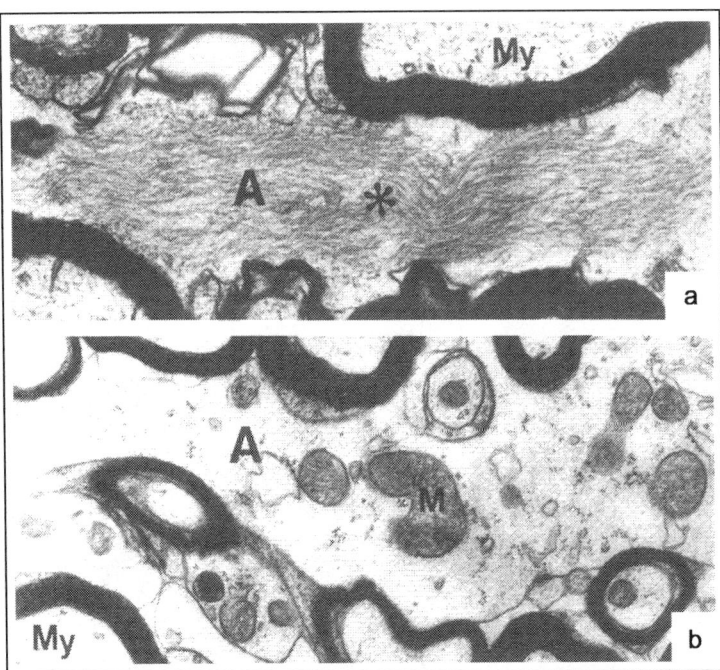

Figure 5. While wild-type astrocytes in the intact CNS contain abundant IFs (asterisk, a), *GFAP$^{-/-}$* astrocytes are devoid of IFs (b). These electron micrographs show cytoplasmic details from astrocytes in the dorsal funiculus of the cervical spinal cord of healthy adult mice. A, astrocytic process; M, mitochondrion; My, myelinated nerve fiber. Reproduced, with permission, from reference 99.

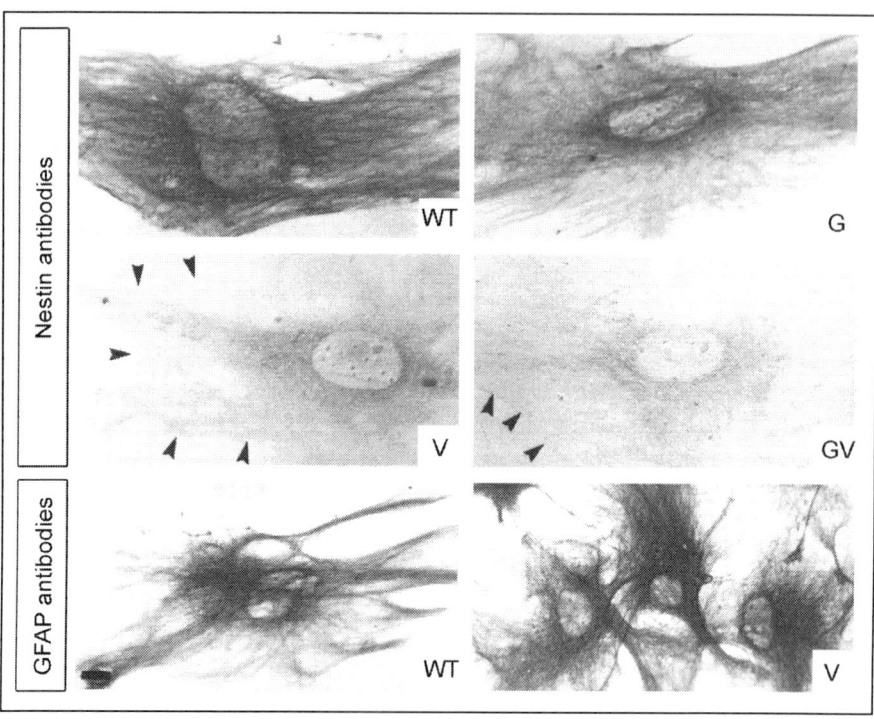

Figure 6. Reactive astrocytes in a primary culture prepared from wild-type (WT), $GFAP^{-/-}$ (G), $Vim^{-/-}$ (V) and $GFAP^{-/-}Vim^{-/-}$ (GV) mice. Nestin antibodies visualize bundles of IFs in wild-type and $GFAP^{-/-}$ astrocytes but fail to do so in $Vim^{-/-}$ or $GFAP^{-/-}Vim^{-/-}$ astrocytes indicating that nestin can neither copolymerize nor coassemble with GFAP. This was also confirmed on a biochemical level.[96] $Vim^{-/-}$ reactive astrocytes contain IF bundles that can be visualized by antibodies against GFAP, albeit with a reduced distance between individual IFs (see Fig. 4 and Table 2). Bar, 10 μm. Reproduced, with permission, from reference 96.

both GFAP and vimentin, suggesting that in astroglial cells, GFAP and vimentin may be necessary for synemin polymerization.[102]

Astrocyte IFs, CNS Trauma and Regeneration

Astrocytes are implicated to be actively involved in many CNS pathologies, such as trauma, ischemia, or neurodegenerative diseases. In response to essentially any CNS pathology, astrocytes undergo a characteristic change in appearance—the hypertrophy of their cellular processes, a phenomenon known as reactive gliosis with characteristic upregulation of IFs (Fig. 7) as well as altered expression profiles of many proteins.[103,104] Thus, genetic depletion of astrocyte IF proteins appeared to be a way to learn more about the physiological and pathological function of astrocytes.[105] Recent data suggest that IFs in astrocytes are structures of major importance in various pathological situations.

Experiments with unchallenged mice deficient for GFAP and/or vimentin did not show major CNS phenotypes,[99,100,106,107] but suggested that astrocytes influence neuronal physiology in the hippocampus[100,108] and in the cerebellum.[109,110] The absence of IF proteins in astroglial cells seems to alter communication between Bergmann glia and Purkinje cells, and it results in impaired eyeblink conditioning and long-term depression in the cerebellum of $GFAP^{-/-}$ mice[109] and impaired motor coordination in $Vim^{-/-}$ mice.[110] It is important to note that one of the four groups that independently generated $GFAP^{-/-}$ mice reported white matter pathologies and dysmyelination in their unchallenged $GFAP^{-/-}$ mice,[111] while the other three groups did not

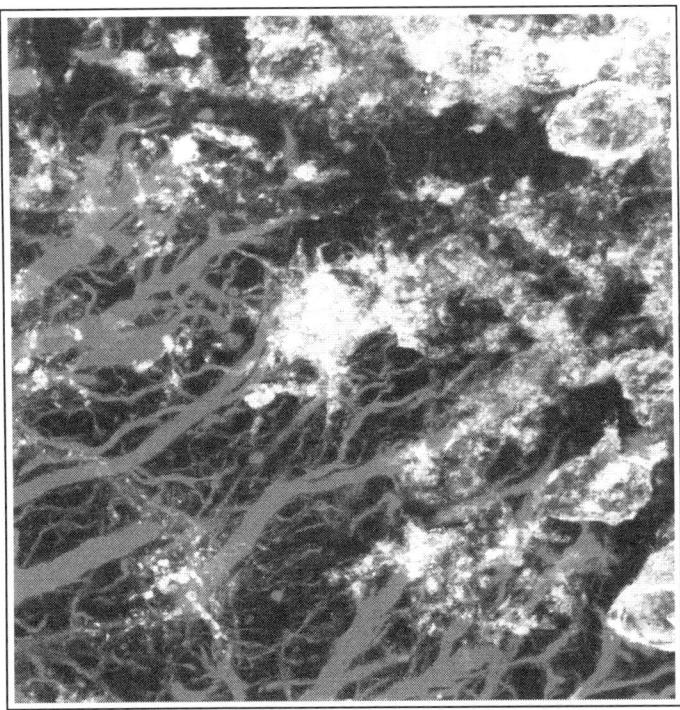

Figure 7. Processes of reactive astrocytes undergo hypertrophy and extend towards the mechanical, ischemic, electric or other type of injury or towards a region affected by neurodegeneration. Here astrocyte processes (gray) were visualized with antibodies against GFAP in a mouse model of electrically induced injury. Note the characteristic boarder between the terminal segments of astrocytic processes and activated microglia (white) visualized by isolectin staining. Reproduced, with permission, from reference 176.

see such changes in their respective *GFAP*[-/-] mice.[99,100,107] This discrepancy has not been resolved. Interestingly, the same group reported increased susceptibility of *GFAP*[-/-] mice to experimental autoimmune encephalomyelitis, a model of multiple sclerosis.[112]

Several trauma models were applied to mice deficient in GFAP and/or vimentin to assess the role of IF upregulation in reactive astrocytes in CNS injury. In one study, we used fine needle injury of the brain cortex and transection of the dorsal funiculus in the upper thoracic spinal cord. The responses of wild-type, *GFAP*[-/-] and in *Vim*[-/-] mice were indistinguishable. In *GFAP*[-/-]*Vim*[-/-] mice, however, the posttraumatic glial scarring was considerably looser and less organized, suggesting that upregulation of IFs is an important step in astrocyte activation and that reactive astrocytes play a role in post-traumatic healing[113] (Fig. 8). Similarly, extended healing period following CNS injury was reported in mice in which dividing astrocytes had been ablated by GFAP-driven expression of herpes simplex virus thymidine kinase and administration of ganciclovir.[114,115]

Hemisection of the spinal cord at T12, reported in another study, was associated with increased axonal sprouting and better functional recovery in *GFAP*[-/-]*Vim*[-/-] mice than wild-type controls.[116] Somewhat conflicting data came out of the work addressing the role of astrocyte IFs in neurite outgrowth in vitro.[98,117,118] One group reported that *GFAP*[-/-]*Vim*[-/-] and *GFAP*[-/-] astrocytes were a better substrate for the outgrowth of neurites in vitro than wild-type astrocytes.[98,118] The other group fou-nd comparable neurite outgrowth when neurons were cultured on wild-type and *GFAP*[-/-] astrocytes.[117] The latter finding is compatible with the normal axonal sprouting and regeneration assessed after dorsal hemisection of the spinal cord in *GFAP*[-/-] mice.[119]

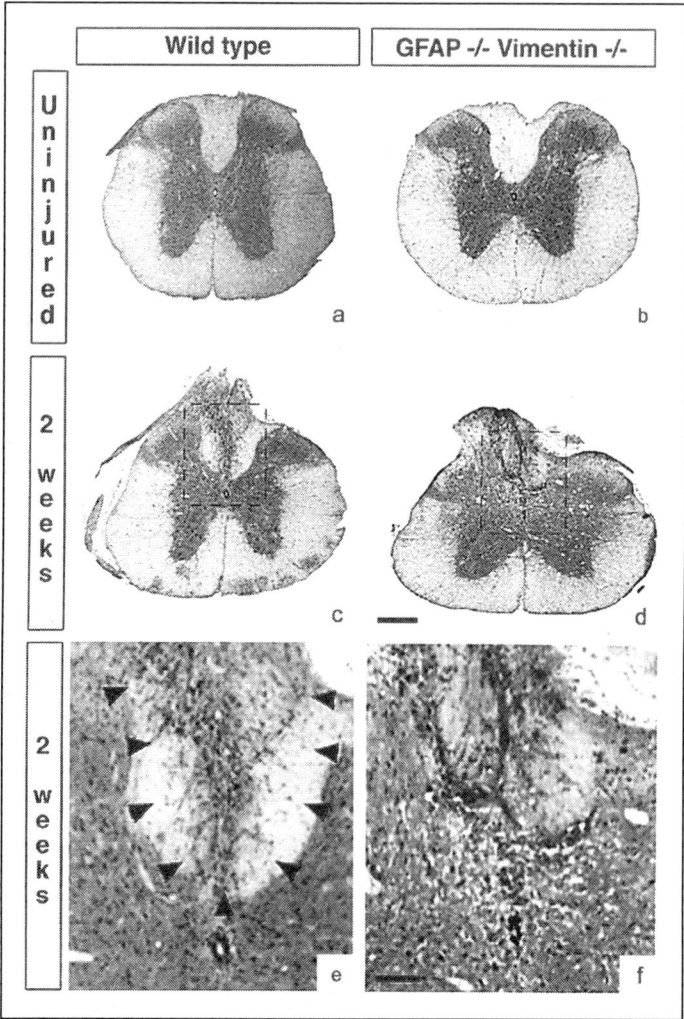

Figure 8. Wound healing after transection of the dorsal funiculus in the upper thoracic spinal cord takes longer in $GFAP^{-/-}Vim^{-/-}$ than wild-type mice, and the resulting glial scarring is reduced. H & E staining. Bar, 300 μm in a-d and 100 μm in e-f. Reproduced from reference 113, ©1999 with permission from The Rockefeller University Press.

In a study conducted in mice with entorhinal cortex lesions, we recently showed that reactive astrocytes devoid of IFs $(GFAP^{-/-}Vim^{-/-})$ exhibited only limited hypertrophy of cell processes. Many processes of $GFAP^{-/-}Vim^{-/-}$ astrocytes were shorter than those of wild-type astrocytes and were not straight, although the volume of the CNS tissue reached by a single astrocyte was comparable to wild-type mice[120] (Fig. 9). In $GFAP^{-/-}Vim^{-/-}$ mice, loss of neuronal synapses in the projection area of the entorhinal cortex (molecular layer of the dentate gyrus of the hippocampus) was prominent 4 days after lesioning (Fig. 10a-b,e), and there was remarkable synaptic regeneration 10 days later (at 14 days after lesions) (Fig. 10c-e). We reported that in contrast to wild-type, $GFAP^{-/-}Vim^{-/-}$ reactive astrocytes did not upregulate their expression of endothelin B receptors, suggesting that the upregulation of this novel marker of reactive astrocytes[121-124] is IF-dependent.[120]

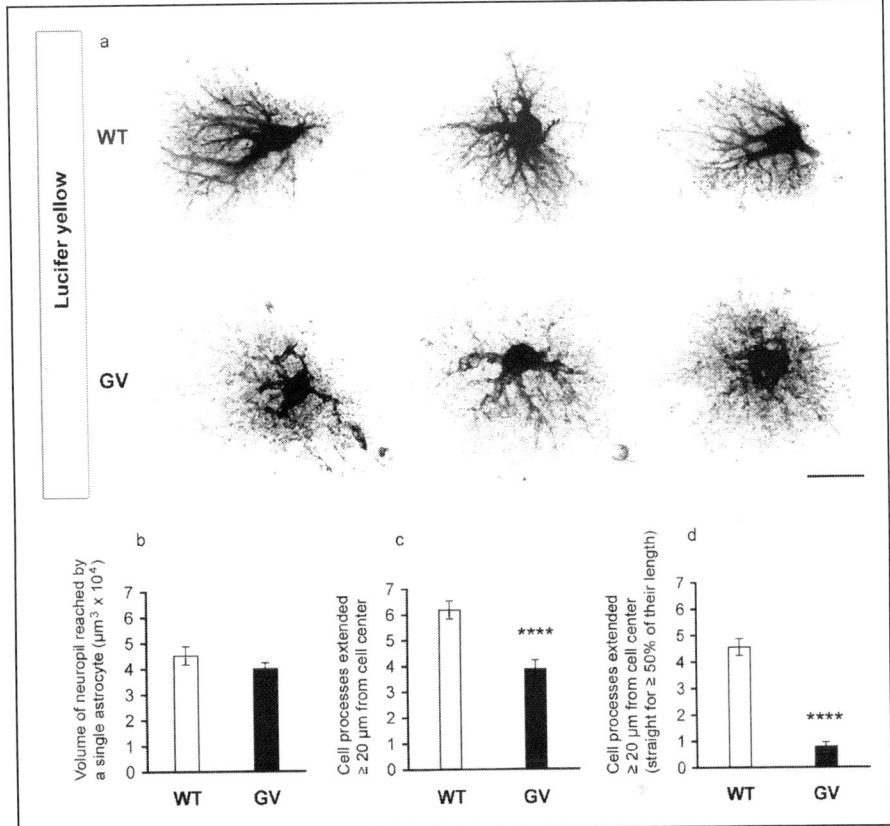

Figure 9. *GFAP$^{-/-}$Vim$^{-/-}$* (GV) reactive astrocytes have fewer long and straight cellular processes than wild-type (WT) as shown by the three-dimensional reconstruction of dye-filled reactive astrocytes in the dentate gyrus of the hippocampus after entorhinal cortex lesions (a,c,d). Wild-type and IF-free *GFAP$^{-/-}$ Vim$^{-/-}$* reactive astrocytes reach comparable volumes of brain tissue (b). ****, p<0.0001; Bar, 20 μm. Reproduced from reference 120, ©2004 with permission from the Society for Neuroscience.

These findings, along with in vitro data on the morphology of IF-depleted astrocytes in primary cultures,[97] showed a novel role for IFs in determining astrocyte morphology. Studies of IF-deficient astrocytes in vitro also implicated the role of astrocyte IF in cell motility. Lepekhin and coworkers assessed the motility of primary cultured astrocytes from *GFAP$^{-/-}$*, *Vim$^{-/-}$*, and *GFAP$^{-/-}$Vim$^{-/-}$* mice and showed that the fast-moving subpopulation was depleted partially among *GFAP$^{-/-}$* and *Vim$^{-/-}$* astrocytes and more severely among *GFAP$^{-/-}$Vim$^{-/-}$* astrocytes[97] (Fig. 11). The in vivo relevance of these findings and the molecular mechanisms involved remain to be established. However, since astrocytes migrate over considerable distances to sites of injury,[125] the slower migration of IF-deficient astrocytes could contributed to the more discrete development of posttraumatic glial scars seen in *GFAP$^{-/-}$Vim$^{-/-}$* mice.[113] The IFs have been implicated in cell motility also in other cells than astrocytes. For example, in vitro studies focusing on the motility of *Vim$^{-/-}$* fibroblasts[126] showed reduced resistance to mechanical stress and reduced migration of these cells in both the scrape wound assay and in Boyden chambers compared to wild-type fibroblasts.[127] Interestingly, another study using monolayer wounding experiments, showed comparable mobility of polarized wild-type and *Vim$^{-/-}$* fibroblasts at the edge of the wound.[128]

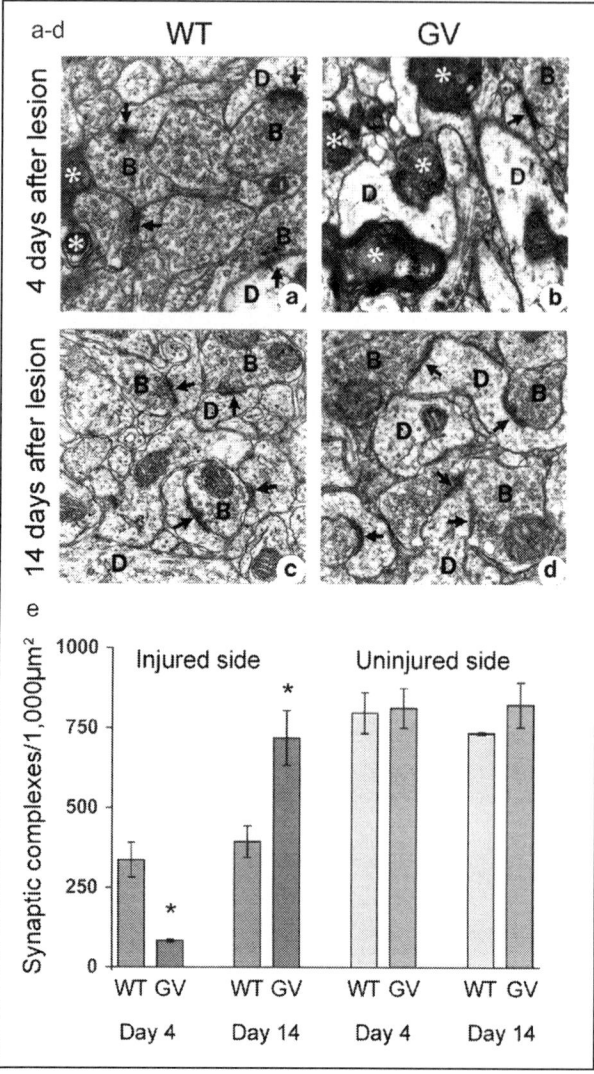

Figure 10. The consequences of lesioning of the entorhinal cortex as seen in its projection area in the dentate gyrus of the hippocampus in *GFAP⁻/⁻Vim⁻/⁻* (GV) and wild-type (WT) mice. At day 4 after lesioning, the synaptic loss and the signs of neurodegeneration were more prominent in *GFAP⁻/⁻Vim⁻/⁻* than wild-type mice (a-b,e). At day 14 after lesioning, the number of synapses in *GFAP⁻/⁻Vim⁻/⁻*, but not wild-type mice, recovered reaching the levels comparable with the uninjured hemisphere (c-e). Asterisks, degenerated axons; arrows, synaptic complexes; D, dendritic profile; B, synaptic bouton; *, p<0.05. Reproduced from reference 120, ©2004 with permission from the Society for Neuroscience.

The emerging picture suggests that the effect of reactive astrocytes after brain or spinal cord trauma is twofold: reactive astrocytes play a beneficial role in the acute stage after CNS injury, but later on act as strong inhibitors of CNS regeneration. These studies of IF-null mutants provided insights into how reactive astrocytes affect the clinical outcome of various CNS pathologies. It is feasible that, by affecting the abundance or the composition of IFs, it might be possible to control the state of cellular differentiation and thus many cellular functions,

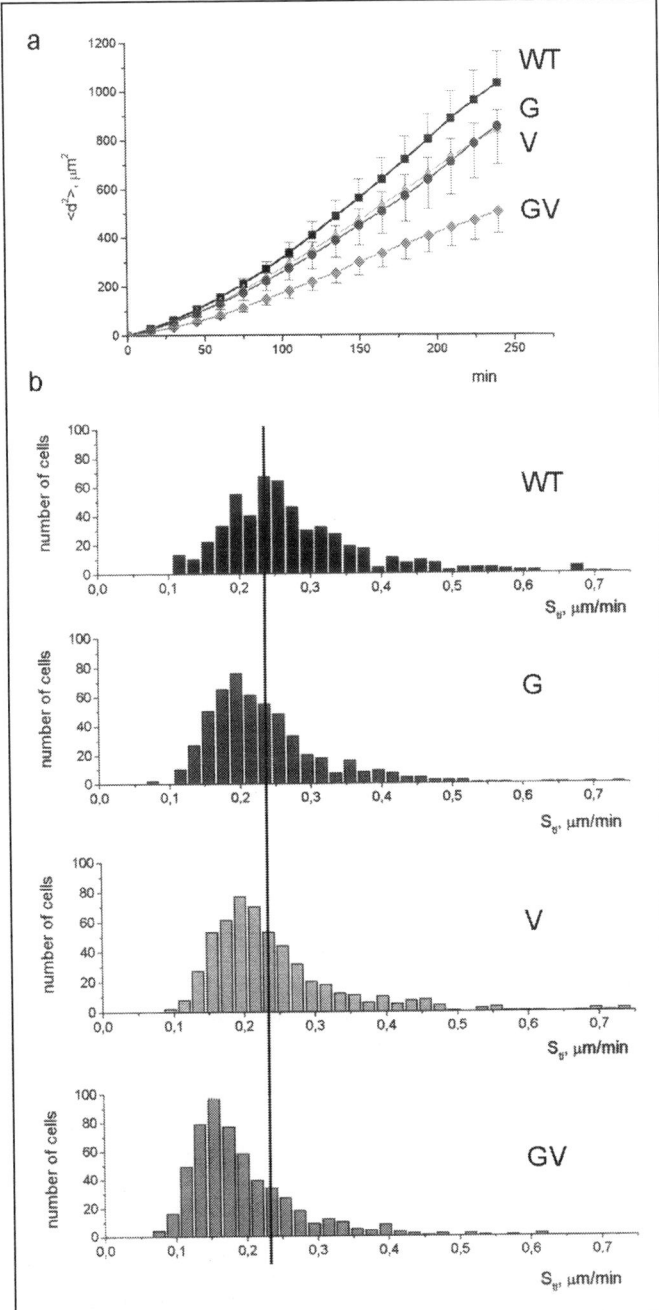

Figure 11. Compared to wild-type, the migration of *GFAP*$^{-/-}$*Vim*$^{-/-}$ reactive astrocytes in vitro is reduced, with the single mutants migrating more slowly than wild-type but faster than *GFAP*$^{-/-}$*Vim*$^{-/-}$ astrocytes (a). Fast-moving subpopulations of *GFAP*$^{-/-}$*Vim*$^{-/-}$ (GV) astrocytes are smaller than in wild-type (WT), with *GFAP*$^{-/-}$ (G) astrocytes and *Vim*$^{-/-}$ (V) astrocytes exhibiting a dose effect (b). Reproduced with permission from reference 97.

which ultimately allow control of complex processes such as the permissiveness of the CNS for regeneration. Such knowledge can be expected to open the way for modulation of astrocyte reactivity for the therapeutic benefit of patients.

Because of their morphology and abundance in the adult CNS (Fig. 1), astrocytes come in direct physical contact with any cell that moves from one place to another. This might be of major importance in situations such as the migration of immature neurons born from endogenous neuronal stem cells, but also for the migration of neuronal precursors from CNS transplants. To assess the impact of astrocyte IFs on the fate of neural transplants, the Chen and Pekny groups transplanted dissociated retinal cells from 0–3-week-old donor mice that ubiquitously express enhanced green fluorescent protein[129] into the retinas of adult wild-type and $GFAP^{-/-}Vim^{-/-}$ recipients and compared the efficiency of integration.[130] In wild-type hosts, few transplanted cells migrated from the transplantation site, and few integrated into the retina. In $GFAP^{-/-}Vim^{-/-}$ hosts, however, the transplanted cells effectively moved through the retina, differentiated into neurons, integrated into the ganglion cell layer, and some of them even extended neurites about 1 mm into the optic nerve (Fig. 12a-d) with the single mutants exhibiting a dose-effect (Fig. 12e-i). Six months after transplantation, the cells remained alive and well integrated $GFAP^{-/-}Vim^{-/-}$ hosts.[130]

Thus, the absence of IFs in astroglial cells of the retina (astrocytes and Müller cells) increases the permissiveness of the retinal environment for integration of neural transplants through the mechanism that is still to be elucidated. The extent to which this reflects increased permissiveness for the migration of transplanted cells remains to be established. However, it is tempting to speculate that IF depletion in astroglial cells alters their differentiation state, rendering them functionally similar to more immature astrocytes, which are also more supportive of CNS regeneration.[131-133] It might be possible that the approaches that would control the expression of IFs and consequently affect cellular differentiation might also be applicable outside the CNS.

Astrocyte IFs and Resistance to Mechanical Stress

One can envisage two functions of astrocytes in a physical trauma inflicted upon the CNS. One is the role in the healing process discussed above, the other might be the ability to increase the mechanical resistance of the CNS and thus making the CNS more resilient to mechanical stress. The abundance of astrocytes throughout the CNS as well as their morphology (Fig. 1) justifies this view. While in other tissues, in particular, the epidermis, the connection between keratin IFs and resistance to mechanical stress is well established (for review see refs. 134,135), the function of astrocyte IFs in maintaining the mechanical integrity of the CNS has been unclear. In $GFAP^{-/-}$ mice, nonreactive astrocytes—which account for the overwhelming majority of astrocytes in a healthy brain—are essentially devoid of IFs.[99,100] Nevertheless, in three independent studies, $GFAP^{-/-}$ mice lived normal lives and, if not challenged, had normal CNS morphology.[99,100,107] Since the CNS is mechanically well protected, the importance of astrocyte IFs for stabilizing the CNS tissue might become manifest only in situations of severe mechanical stress. Two series of experiments using the head percussion model and the severe mechanical stress applied on the retina, respectively, suggested that this indeed is the case.

$GFAP^{-/-}$ mice were subjected to head injury from a dropped weight. When placed on a wooden board to prevent head movement at impact, $GFAP^{-/-}$ mice survived as well as wild-type controls. However, when placed on a foam bed that allowed head movement at impact, most of the $GFAP^{-/-}$ mice, but none of the wild-type controls, died after the injury. The $GFAP^{-/-}$ mice showed prominent subpial and white matter bleeding in the region of the cervical spinal cord, possibly resulting from a vein rupture.[136]

We recently assessed the effect of the absence of IFs in astroglial cells on the mechanical stability of the retina under severe mechanical stress. Being an accessible part of the CNS, retina is well suited for such experiments, in this case performed in mice just seconds after death while the retinal tissue was still alive. Application of a severe mechanical stress left the retinas of wild-type controls intact. However, in $GFAP^{-/-}Vim^{-/-}$ mice and, to a lesser extent, in $Vim^{-/-}$ mice, the inner limiting membrane and adjacent tissue separated from the rest of the retina (Fig. 13). Electron microscopy showed that this retinal "crack" occurred within

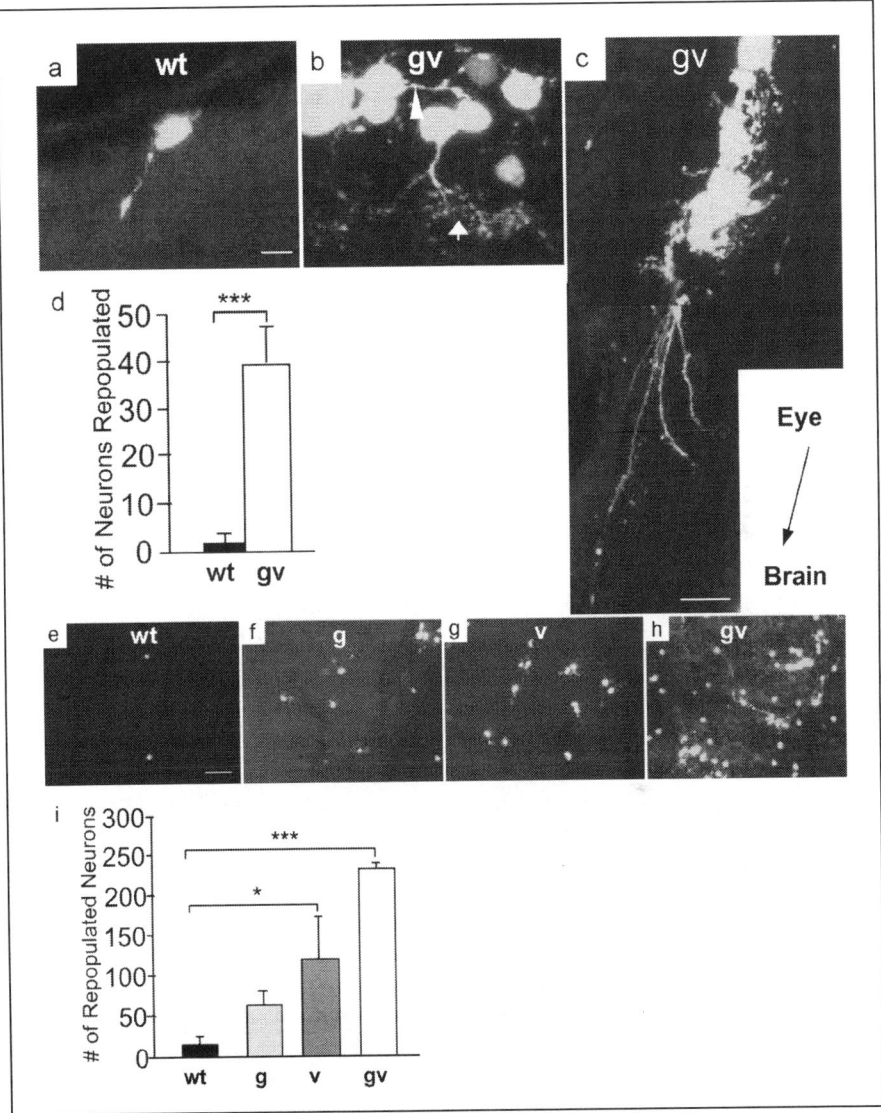

Figure 12. Retinal transplants from mice ubiquitously expressing enhanced green fluorescent protein integrated much better in $GFAP^{-/-}Vim^{-/-}$ (gv) than wild-type (wt) recipients (a-d). In $GFAP^{-/-}Vim^{-/-}$ recipients, transplanted cells migrated more efficiently from the transplantation site and integrated into the ganglion cell layer (GCL, d), exhibiting typical morphology of ganglion cells with axon-like process parallel to the retinal surface (arrowhead) and branched dendritic tree-like structures (arrow, b). Some of these neurons even extended axons into the optic nerve (c). In single mutant recipients (g or v), the transplanted cells spread out more extensively than in wild-type but less efficiently than in $GFAP^{-/-}Vim^{-/-}$ recipients (e-i). *, p<0.05; ***, p<0.001. Bar, 5 μm in a-b, 50 μm in c, 100 μm in e-h. Data represent mean ± SD. Reproduced, with permission, from reference 130.

the endfeet of Müller cells—radial glia-like astroglial cells in the retina that normally contain IFs composed of GFAP and vimentin.[137] Thus, at least in two specific regions of the CNS, astrocyte IFs seem to be important for resistance to severe mechanical stress.

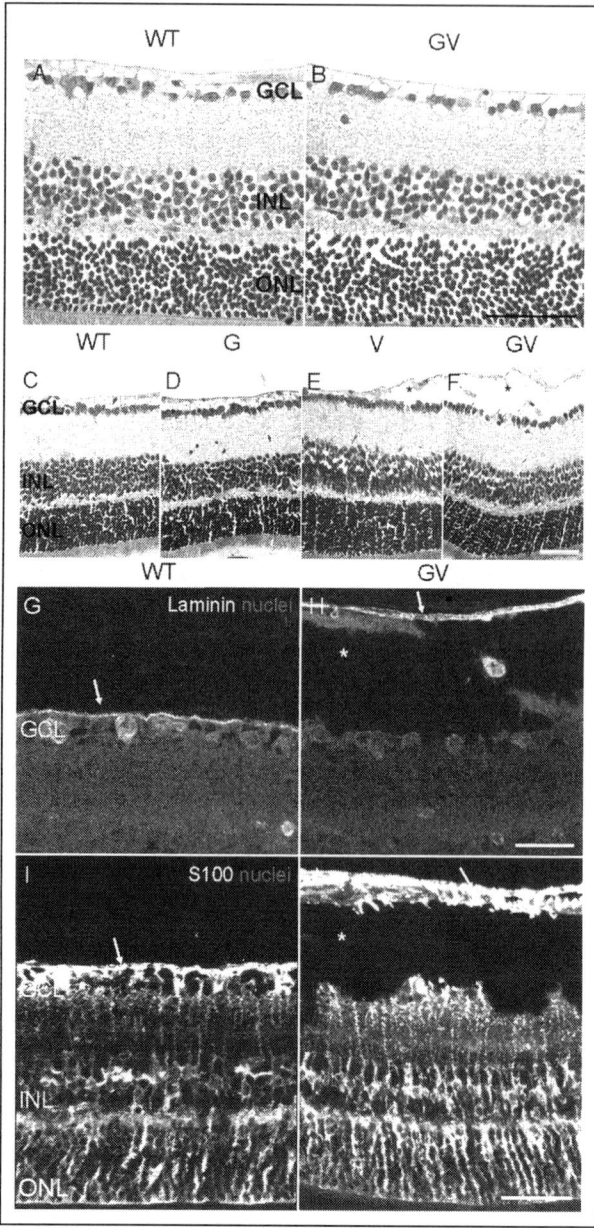

Figure 13. *GFAP$^{-/-}$Vim$^{-/-}$* and wild-type retinas are indistinguishable in the absence of a major mechanical challenge (A,B). Severe mechanical stress on the retina leads to the complete and partial separation of the inner limiting membrane and adjacent tissue from the rest of the retina (asterisk) in *GFAP$^{-/-}$Vim$^{-/-}$* (GV) and partial separation (asterisk) in *Vim$^{-/-}$* (V) mice (E,F,H,J). The retinas of wild-type (WT; C,G,I) or *GFAP$^{-/-}$* (G; D) mice remain intact. H & E staining (A-F). Visualization of the inner limiting membrane by antibodies against laminin (G,H) and of Müller cells and astrocytes by antibodies against S-100 (I,J). GCL, ganglion cell layer; INL, inner nuclear layer; ONL, outer nuclear layer; arrow, the inner limiting membrane. Reproduced, with permission, from reference 137.

Astrocyte IFs, Osmotic Stress and Ischemia/Hypoxia

In vitro, astrocytes respond to a hypoosmotic environment by transient swelling and within minutes show a tendency to return to their original cell volume.[138,139] This phenomenon, known as regulatory volume decrease, involves an efflux of osmotically active molecules from astrocytes, such as the amino acid taurine.[138,140-142] It was proposed that regulatory volume decrease by astrocytes might be the key mechanism in counteracting the development of brain edema in response to brain ischemia or trauma and that cytoskeleton-linked stretch-activated plasma membrane channels serve as cell-volume sensors.[143-146]

Ding and coworkers subjected primary astrocyte cultures from wild-type, $GFAP^{-/-}$, $Vim^{-/-}$, and $GFAP^{-/-}Vim^{-/-}$ mice to hypoosmotic stress (corresponding to a 25 mM reduction in NaCl) in perfusion chambers and assessed the efflux of ^{3}H-taurine. Taurine release was up to 50% lower in $GFAP^{-/-}Vim^{-/-}$ than wild-type astrocytes, but tended to be only slightly decreased in the single mutants.[147] Anderova and coworkers perfused spinal slices with an isoosmotic solution with an increased concentration of potassium (50 mM) or a hypoosmotic solution with a reduced sodium concentration and found smaller increases in the potassium concentration around astrocytes in slices from $GFAP^{-/-}$ mice than in those from wild-type controls.[148] Thus, genetic ablation of astrocytic IFs seems to compromise the ability of astrocytes to respond to hypoosmotic stress.

Do these findings have any relevance for brain pathologies, in particular those connected with prominent osmotic stress, such as brain ischemia? Nawashiro and coworkers exposed $GFAP^{-/-}$ and wild-type mice to brain ischemia induced by middle cerebral artery occlusion for 2 days and reported comparable infarct volumes in the two groups. However, when middle cerebral artery occlusion was combined with transient occlusion of the carotid artery, $GFAP^{-/-}$ mice had larger infarcts than controls.[136] This raises the interesting and unresolved question of whether reactive astrocytes protect ischemically compromised brain tissue around the infarct in stroke patients. In this respect, $GFAP^{-/-}$ astrocytes in culture showed increased intracellular glutamine levels[149] and decreased glutamate transport.[150] Studies of $GFAP^{-/-}Vim^{-/-}$ mice, whose reactive astrocytes are devoid of IFs,[96] in various brain ischemia paradigms should shed more light on this issue.

We have recently turned to a hypoxia model in order to assess the pathophysiological implications of the "retinal crack" phenotype found in $GFAP^{-/-}Vim^{-/-}$ mice described above. We exposed $GFAP^{-/-}Vim^{-/-}$ mice and single mutants to retinal hypoxia. This leads to oxygen-induced retinopathy, a widely used model of retinopathy of immaturity that also exhibits some features of diabetic retinopathy.[151] On postnatal day 7, mice are placed into an environment with decreased oxygen concentration, which delays the development of the vascular system. Five days later, the mice are transferred to normooxygenic environment, which leads to massive neovascularization triggered by relative hypoxia (Fig. 14). The vessels grow from the retina into the vitreous body (as they do in premature babies or patients with diabetes), and their presence there can easily be quantified.[151] Hypoxia-induced vascularization was decreased substantially in $GFAP^{-/-}Vim^{-/-}$ and partially in $Vim^{-/-}$ mice (Fig. 14). Thus, the absence of IFs in Müller cells of the retina decreases the resistance of their endfeet and consequently of the corresponding layer of the retina to mechanical stress, and it also reduces the extent of ischemia-triggered pathological vascularization.[137]

Astrocyte IFs, Cell Proliferation and Tumorigenesis

High grade astrocytomas are the most frequent brain tumors, and they are ranked among the most malignant tumors.[152] In many high grade astrocytomas, the tumor cells lose their GFAP expression, and this was often inversely correlated with the malignancy of the tumor.[153-157] Within the same tumor, the cells negative for GFAP were shown to grow faster than the GFAP-positive cells in their vicinity.[158,159] These findings were compatible with a number of in vitro studies. The inhibition of GFAP expression by anti-sense cDNA in human astrocytoma cell lines resulted in both the increased cell proliferation, transformability and the loss of the

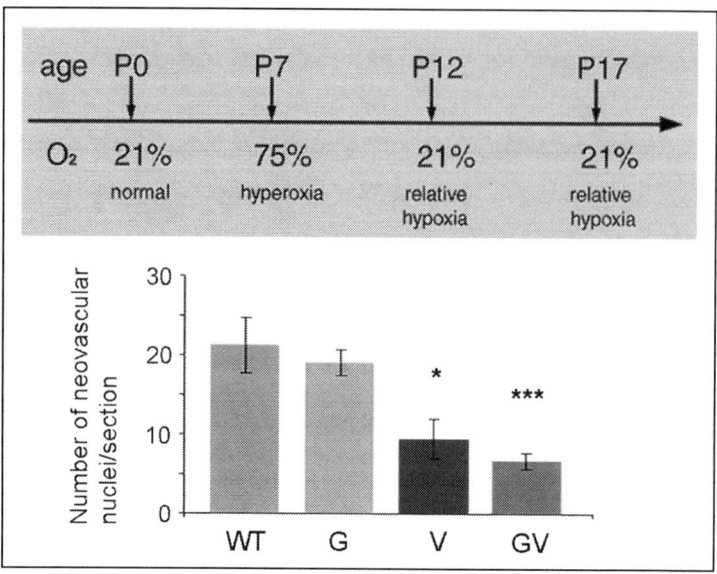

Figure 14. A schematic presentation of the oxygen-induced retinopathy model. The number of neovascular nuclei in the vitreous body, a measure of the extent of hypoxia-induced pathological vascularization at postnatal day 17 (P17), is substantially decreased in $GFAP^{-/-}Vim^{-/-}$ (GV) retinas and modestly decreased in $Vim^{-/-}$ (V) retinas. No difference was found between $GFAP^{-/-}$ (G) and wild-type (WT) retinas. In the absence of hypoxia, blood vessels do not enter the vitreous body, and the normal vascularization of the retina does not depend on the presence of GFAP and vimentin. *, p<0.05; ***, p<0.005. Reproduced with permission from reference 137.

tumor cells to extend processes in response to neurons.[160,161] Restoration of GFAP expression reinduced process extension in the tumor cells.[162] Transfections of a GFAP-negative astrocytoma cell-line with GFAP-expressing construct resulted in a decreased proliferation and cell transformability.[163,164] We showed that primary astrocytes from $GFAP^{-/-}$ mice grew more quickly in culture and reached higher saturation cell densities than wild-type cells.[95] To address a possible role of GFAP in tumor development, we induced astrocytomas in $GFAP^{-/-}$ and $GFAP^{+/+}$ mice by prenatal exposure to the mutagen ethylnitrosourea (ENU) on the $p53^{-/-}$ genetic background[165,166] (Fig. 15a). We found no difference in tumor incidence, age at tumor detection, tumor size, location or histology between $GFAP^{-/-}$ and $GFAP^{+/+}$ mice[167] (Fig. 15b-c). Thus, the loss of GFAP expression does not seem to constitute a step in the development of high grade astrocytomas. Most likely, it reflects the undifferentiated state of these cells.[167]

GFAP Mutations and Alexander Disease

Messing, Brenner and coworkers generated mice overexpressing human GFAP in order to study the role of GFAP in astrocyte hypertrophy. The astrocytes of these transgenic mice formed complex intracytoplasmic aggregates of GFAP and small stress proteins that were identical to structures known as Rosenthal fibers.[168] Rosenthal fibers, which accompany chronic reactive astrogliosis, are eosinophilic, elongated structures that, when examined ultrastructurally, appear as electron-dense, amorphous masses surrounded by and merging with dense bundles of IFs[168] (Fig. 16). Rosenthal fibers are also a hallmark of Alexander disease, a rare and fatal leukoencephalopathy that most commonly affects infants and young children, who typically present with feeding problems, paraparesis, seizures and mental and physical retardation. Juvenile forms of Alexander disease cause predominantly pseudobulbar and bulbar signs, while adult forms are more variable and often resemble multiple sclerosis.

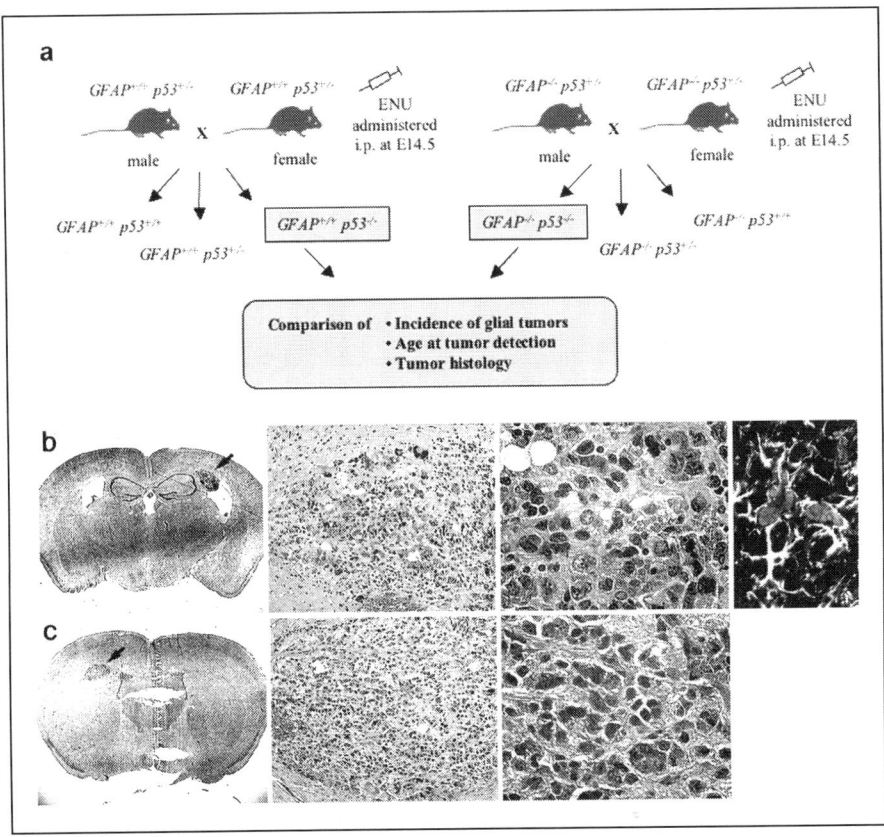

Figure 15. High-grade astrocytomas in *GFAP⁻ᐟ⁻* and *GFAP⁺ᐟ⁺* mice. a) schematic presentation of astrocytoma induction, which was achieved by combining prenatal exposure to ethylnitrosourea (ENU) and genetic absence of both copies of the tumor suppressor gene *p53*. b-c) astrocytomas occured with comparable frequency in *GFAP⁻ᐟ⁻* and *GFAP⁺ᐟ⁺* mice (41 and 34% respectively, within 20 weeks after the induction) and they showed similar histology and progression. Note the abundance of GFAP-positive cells in the tumors in *GFAP⁺ᐟ⁺* mice as visualized by antibodies against GFAP (the panel on the right in b). H & E staining, arrows depict tumors on frontal sections through the brain. Reproduced, with permission, from reference 167.

The GFAP-over-expressing mice died at an early age and although the cause of death remains unknown, these results pointed to an interesting possibility of GFAP as a candidate gene for Alexander disease. And indeed, subsequent investigations determined that the majority of cases of infantile Alexander disease, and at least some cases of the later-onset juvenile and adult forms, are due to heterozygous missense mutations in the GFAP gene. The heterozygosity of the mutations suggests that they are dominant. In the majority of cases, the mutations seemed to occur de novo and were not found in either parent.[169,170] However, familial adult cases have been described, raising the interesting issue of reduced penetrance or germline mosaicism.[171,172] Although these results identify mutated GFAP as at least one of the culprits responsible for a fatal neurological disorder in humans, the mechanism remains unclear. How the mutant GFAP protein causes brain damage and the role of Rosenthal fibers in this process are unknown.[173,174] However, Alexander disease became the first monogenic disease caused by a primary defect in astrocytes and it can be expected that the understanding of its molecular pathogenesis will provide data relevant for other neurodegenerative diseases.

Figure 16. Rosenthal fibers as visualized by transmission electron microscopy with IFs and electron-dense amorphous masses as their characteristic components. Reproduced, with permission, from reference 177.

References

1. Gotz M, Hartfuss E, Malatesta P. Radial glial cells as neuronal precursors: A new perspective on the correlation of morphology and lineage restriction in the developing cerebral cortex of mice. Brain Res Bull 2002; 57(6):777-788.
2. Chanas-Sacre G, Rogister B, Moonen G et al. Radial glia phenotype: Origin, regulation, and transdifferentiation. J Neurosci Res 2000; 61(4):357-363.
3. Alves JA, Barone P, Engelender S et al. Initial stages of radial glia astrocytic transformation in the early postnatal anterior subventricular zone. J Neurobiol 2002; 52(3):251-265.
4. deAzevedo LC, Fallet C, Moura-Neto V et al. Cortical radial glial cells in human fetuses: Depth-correlated transformation into astrocytes. J Neurobiol 2003; 55(3):288-298.
5. Choi BH, Lapham LW. Evolution of Bergmann glia in developing human fetal cerebellum: A Golgi, electron microscopic and immunofluorescent study. Brain Res 1980; 190(2):369-383.
6. Edwards MA, Yamamoto M, Caviness Jr VS. Organization of radial glia and related cells in the developing murine CNS. An analysis based upon a new monoclonal antibody marker. Neuroscience 1990; 36(1):121-144.
7. Dahlstrand J, Lardelli M, Lendahl U. Nestin mRNA expression correlates with the central nervous system progenitor cell state in many, but not all, regions of developing central nervous system. Brain Res Dev Brain Res 1995; 84(1):109-129.
8. Pixley SK, de Vellis J. Transition between immature radial glia and mature astrocytes studied with a monoclonal antibody to vimentin. Brain Res 1984; 317(2):201-209.
9. Kalman M, Szekely AD, Csillag A. Distribution of glial fibrillary acidic protein and vimentin-immunopositive elements in the developing chicken brain from hatch to adulthood. Anat Embryol (Berl) 1998; 198(3):213-235.
10. Levitt P, Rakic P. Immunoperoxidase localization of glial fibrillary acidic protein in radial glial cells and astrocytes of the developing rhesus monkey brain. J Comp Neurol 1980; 193(3):815-840.
11. Sancho-Tello M, Valles S, Montoliu C et al. Developmental pattern of GFAP and vimentin gene expression in rat brain and in radial glial cultures. Glia 1995; 15(2):157-166.

12. Tardy M, Fages C, Riol H et al. Developmental expression of the glial fibrillary acidic protein mRNA in the central nervous system and in cultured astrocytes. J Neurochem 1989; 52(1):162-167.
13. Riol H, Fages C, Tardy M. Transcriptional regulation of glial fibrillary acidic protein (GFAP)-mRNA expression during postnatal development of mouse brain. J Neurosci Res 1992; 32(1):79-85.
14. Sultana S, Sernett SW, Bellin RM et al. Intermediate filament protein synemin is transiently expressed in a subset of astrocytes during development. Glia 2000; 30(2):143-153.
15. Kalman M, Ajtai BM. A comparison of intermediate filament markers for presumptive astroglia in the developing rat neocortex: Immunostaining against nestin reveals more detail, than GFAP or vimentin. Int J Dev Neurosci 2001; 19(1):101-108.
16. Pixley SK, Kobayashi Y, de Vellis J. A monoclonal antibody against vimentin: Characterization. Brain Res 1984; 317(2):185-199.
17. Shaw G, Osborn M, Weber K. An immunofluorescence microscopical study of the neurofilament triplet proteins, vimentin and glial fibrillary acidic protein within the adult rat brain. Eur J Cell Biol 1981; 26(1):68-82.
18. Bushong EA, Martone ME, Jones YZ et al. Protoplasmic astrocytes in CA1 stratum radiatum occupy separate anatomical domains. J Neurosci 2002; 22(1):183-192.
19. Bushong EA, Martone ME, Ellisman MH. Maturation of astrocyte morphology and the establishment of astrocyte domains during postnatal hippocampal development. Int J Dev Neurosci 2004; 22(2):73-86.
20. Song H, Stevens CF, Gage FH. Astroglia induce neurogenesis from adult neural stem cells. Nature 2002; 417(6884):39-44.
21. Morshead CM, Garcia AD, Sofroniew MV et al. The ablation of glial fibrillary acidic protein-positive cells from the adult central nervous system results in the loss of forebrain neural stem cells but not retinal stem cells. Eur J Neurosci 2003; 18(1):76-84.
22. Doetsch F, Caille I, Lim DA et al. Subventricular zone astrocytes are neural stem cells in the adult mammalian brain. Cell 1999; 97(6):703-716.
23. Imura T, Kornblum HI, Sofroniew MV. The predominant neural stem cell isolated from postnatal and adult forebrain but not early embryonic forebrain expresses GFAP. J Neurosci 2003; 23(7):2824-2832.
24. Laywell ED, Rakic P, Kukekov VG et al. Identification of a multipotent astrocytic stem cell in the immature and adult mouse brain. Proc Natl Acad Sci USA 2000; 97(25):13883-13888.
25. Eng LF, Vanderhaeghen JJ, Bignami A et al. An acidic protein isolated from fibrous astrocytes. Brain Res 1971; 28(2):351-354.
26. Eng LF, Ghirnikar RS, Lee YL. Glial fibrillary acidic protein: GFAP-thirty-one years (1969-2000). Neurochem Res 2000; 25(9-10):1439-1451.
27. Reeves SA, Helman LJ, Allison A et al. Molecular cloning and primary structure of human glial fibrillary acidic protein. Proc Natl Acad Sci USA 1989; 86(13):5178-5182.
28. Bernier L, Colman DR, D'Eustachio P. Chromosomal locations of genes encoding 2',3' cyclic nucleotide 3'-phosphodiesterase and glial fibrillary acidic protein in the mouse. J Neurosci Res 1988; 20(4):497-504.
29. Nielsen AL, Jorgensen AL. Structural and functional characterization of the zebrafish gene for glial fibrillary acidic protein, GFAP. Gene 2003; 310:123-132.
30. Balcarek JM, Cowan NJ. Structure of the mouse glial fibrillary acidic protein gene: Implications for the evolution of the intermediate filament multigene family. Nucleic Acids Res 1985; 13(15):5527-5543.
31. Miura M, Tamura T, Mikoshiba K. Cell-specific expression of the mouse glial fibrillary acidic protein gene: Identification of the cis- and trans-acting promoter elements for astrocyte-specific expression. J Neurochem 1990; 55(4):1180-1188.
32. Besnard F, Brenner M, Nakatani Y et al. Multiple interacting sites regulate astrocyte-specific transcription of the human gene for glial fibrillary acidic protein. J Biol Chem 1991; 266(28):18877-18883.
33. Sarid J. Identification of a cis-acting positive regulatory element of the glial fibrillary acidic protein gene. J Neurosci Res 1991; 28(2):217-228.
34. Sarkar S, Cowan NJ. Regulation of expression of glial filament acidic protein. J Cell Sci Suppl 1991; 15:97-102.
35. Kaneko R, Sueoka N. Tissue-specific versus cell type-specific expression of the glial fibrillary acidic protein. Proc Natl Acad Sci USA 1993; 90(10):4698-4702.
36. Brenner M. Structure and transcriptional regulation of the GFAP gene. Brain Pathol 1994; 4(3):245-257.
37. Brenner M, Messing A. GFAP Transgenic Mice. Methods 1996; 10(3):351-364.

38. Johnson WB, Ruppe MD, Rockenstein EM et al. Indicator expression directed by regulatory sequences of the glial fibrillary acidic protein (GFAP) gene: In vivo comparison of distinct GFAP-lacZ transgenes. Glia 1995; 13(3):174-184.
39. Brenner M, Kisseberth WC, Su Y et al. GFAP promoter directs astrocyte-specific expression in transgenic mice. J Neurosci 1994; 14(3 Pt 1):1030-1037.
40. Su M, Hu H, Lee Y et al. Expression specificity of GFAP transgenes. Neurochem Res 2004; 29:2075-2093.
41. Masood K, Besnard F, Su Y et al. Analysis of a segment of the human glial fibrillary acidic protein gene that directs astrocyte-specific transcription. J Neurochem 1993; 61(1):160-166.
42. Yu AC, Lee YL, Fu WY et al. Gene expression in astrocytes during and after ischemia. Prog Brain Res 1995; 105:245-253.
43. Teter B, Osterburg HH, Anderson CP et al. Methylation of the rat glial fibrillary acidic protein gene shows tissue-specific domains. J Neurosci Res 1994; 39(6):680-693.
44. Condorelli DF, Dell'Albani P, Conticello SG et al. A neural-specific hypomethylated domain in the 5' flanking region of the glial fibrillary acidic protein gene. Dev Neurosci 1997; 19(5):446-456.
45. Barresi V, Condorelli DF, Giuffrida Stella AM. GFAP gene methylation in different neural cell types from rat brain. Int J Dev Neurosci 1999; 17(8):821-828.
46. Condorelli DF, Nicoletti VG, Barresi V et al. Tissue-specific DNA methylation patterns of the rat glial fibrillary acidic protein gene. J Neurosci Res 1994; 39(6):694-707.
47. Teter B, Rozovsky I, Krohn K et al. Methylation of the glial fibrillary acidic protein gene shows novel biphasic changes during brain development. Glia 1996; 17(3):195-205.
48. Takizawa T, Nakashima K, Namihira M et al. DNA methylation is a critical cell-intrinsic determinant of astrocyte differentiation in the fetal brain. Dev Cell 2001; 1(6):749-758.
49. Song MR, Ghosh A. FGF2-induced chromatin remodeling regulates CNTF-mediated gene expression and astrocyte differentiation. Nat Neurosci 2004; 7(3):229-235.
50. Bernal SD, Stahel RA. Cytoskeleton-associated proteins: Their role as cellular integrators in the neoplastic process. Crit Rev Oncol Hematol 1985; 3(3):191-204.
51. Kachinsky AM, Dominov JA, Miller JB. Myogenesis and the intermediate filament protein, nestin. Dev Biol 1994; 165(1):216-228.
52. Sejersen T, Lendahl U. Transient expression of the intermediate filament nestin during skeletal muscle development. J Cell Sci 1993; 106(Pt 4):1291-1300.
53. Frisen J, Johansson CB, Torok C et al. Rapid, widespread, and longlasting induction of nestin contributes to the generation of glial scar tissue after CNS injury. J Cell Biol 1995; 131(2):453-464.
54. Sugawara K, Kurihara H, Negishi M et al. Nestin as a marker for proliferative endothelium in gliomas. Lab Invest 2002; 82(3):345-351.
55. Paulin D, Lilienbaum A, Duprey P et al. Regulatory elements of the human vimentin gene: Activation during proliferation. Reprod Nutr Dev 1990; 30(3):423-429.
56. Izmailova ES, Wieczorek E, Perkins EB et al. A GC-box is required for expression of the human vimentin gene. Gene 1999; 235(1-2):69-75.
57. Kryszke MH, Vicart P. Regulation of the expression of the human vimentin gene: Application to cellular immortalization. Pathol Biol (Paris) 1998; 46(1):39-45.
58. Pieper FR, Slobbe RL, Ramaekers FC et al. Upstream regions of the hamster desmin and vimentin genes regulate expression during in vitro myogenesis. EMBO J 1987; 6(12):3611-3618.
59. Sommers CL, Skerker JM, Chrysogelos SA et al. Regulation of vimentin gene transcription in human breast cancer cell lines. Cell Growth Differ 1994; 5(8):839-846.
60. Moura-Neto V, Kryszke MH, Li Z et al. A 28-bp negative element with multiple factor-binding activity controls expression of the vimentin-encoding gene. Gene 1996; 168(2):261-266.
61. Wieczorek E, Lin Z, Perkins EB et al. The zinc finger repressor, ZBP-89, binds to the silencer element of the human vimentin gene and complexes with the transcriptional activator, Sp1. J Biol Chem 2000; 275(17):12879-12888.
62. Zhang X, Diab IH, Zehner ZE. ZBP-89 represses vimentin gene transcription by interacting with the transcriptional activator, Sp1. Nucleic Acids Res 2003; 31(11):2900-2914.
63. van de Klundert FA, van Eldik GJ, Pieper FR et al. Identification of two silencers flanking an AP-1 enhancer in the vimentin promoter. Gene 1992; 122(2):337-343.
64. Pieper FR, Van de Klundert FA, Raats JM et al. Regulation of vimentin expression in cultured epithelial cells. Eur J Biochem 1992; 210(2):509-519.
65. Lilienbaum A, Paulin D. Activation of the human vimentin gene by the Tax human T-cell leukemia virus. I. Mechanisms of regulation by the NF-kappa B transcription factor. J Biol Chem 1993; 268(3):2180-2188.
66. Josephson R, Muller T, Pickel J et al. POU transcription factors control expression of CNS stem cell-specific genes. Development 1998; 125(16):3087-3100.

67. Lothian C, Lendahl U. An evolutionarily conserved region in the second intron of the human nestin gene directs gene expression to CNS progenitor cells and to early neural crest cells. Eur J Neurosci 1997; 9(3):452-462.
68. Lothian C, Prakash N, Lendahl U et al. Identification of both general and region-specific embryonic CNS enhancer elements in the nestin promoter. Exp Cell Res 1999; 248(2):509-519.
69. Yaworsky PJ, Kappen C. Heterogeneity of neural progenitor cells revealed by enhancers in the nestin gene. Dev Biol 1999; 205(2):309-321.
70. Zimmerman L, Parr B, Lendahl U et al. Independent regulatory elements in the nestin gene direct transgene expression to neural stem cells or muscle precursors. Neuron 1994; 12(1):11-24.
71. Johansson CB, Lothian C, Molin M et al. Nestin enhancer requirements for expression in normal and injured adult CNS. J Neurosci Res 2002; 69(6):784-794.
72. Rueger DC, Huston JS, Dahl D et al. Formation of 100 A filaments from purified glial fibrillary acidic protein in vitro. J Mol Biol 1979; 135(1):53-68.
73. Renner W, Franke WW, Schmid E et al. Reconstitution of intermediate-sized filaments from denatured monomeric vimentin. J Mol Biol 1981; 149(2):285-306.
74. Angelides KJ, Smith KE, Takeda M. Assembly and exchange of intermediate filament proteins of neurons: Neurofilaments are dynamic structures. J Cell Biol 1989; 108(4):1495-1506.
75. Nakamura Y, Takeda M, Angelides KJ et al. Assembly, disassembly, and exchange of glial fibrillary acidic protein. Glia 1991; 4(1):101-110.
76. Miller RK, Vikstrom K, Goldman RD. Keratin incorporation into intermediate filament networks is a rapid process. J Cell Biol 1991; 113(4):843-855.
77. Wiegers W, Honer B, Traub P. Microinjection of intermediate filament proteins into living cells with and without preexisting intermediate filament network. Cell Biol Int Rep 1991; 15(4):287-296.
78. Vikstrom KL, Lim SS, Goldman RD et al. Steady state dynamics of intermediate filament networks. J Cell Biol 1992; 118(1):121-129.
79. Yoon M, Moir RD, Prahlad V et al. Motile properties of vimentin intermediate filament networks in living cells. J Cell Biol 1998; 143(1):147-157.
80. Goldman RD, Chou YH, Prahlad V et al. Intermediate filaments: Dynamic processes regulating their assembly, motility, and interactions with other cytoskeletal systems. FASEB J 1999; 13(Suppl 2):S261-265.
81. Prahlad V, Yoon M, Moir RD et al. Rapid movements of vimentin on microtubule tracks: Kinesin-dependent assembly of intermediate filament networks. J Cell Biol 1998; 143(1):159-170.
82. Helfand BT, Loomis P, Yoon M et al. Rapid transport of neural intermediate filament protein. J Cell Sci 2003; 116(Pt 11):2345-2359.
83. Chou YH, Helfand BT, Goldman RD. New horizons in cytoskeletal dynamics: Transport of intermediate filaments along microtubule tracks. Curr Opin Cell Biol 2001; 13(1):106-109.
84. Inagaki M, Nakamura Y, Takeda M et al. Glial fibrillary acidic protein: Dynamic property and regulation by phosphorylation. Brain Pathol 1994; 4(3):239-243.
85. Inagaki M, Gonda Y, Nishizawa K et al. Phosphorylation sites linked to glial filament disassembly in vitro locate in a nonalpha-helical head domain. J Biol Chem 1990; 265(8):4722-4729.
86. Nishizawa K, Yano T, Shibata M et al. Specific localization of phosphointermediate filament protein in the constricted area of dividing cells. J Biol Chem 1991; 266(5):3074-3079.
87. Nakamura Y, Takeda M, Aimoto S et al. Assembly regulatory domain of glial fibrillary acidic protein. A single phosphorylation diminishes its assembly-accelerating property. J Biol Chem 1992; 267(32):23269-23274.
88. Matsuoka Y, Nishizawa K, Yano T et al. Two different protein kinases act on a different time schedule as glial filament kinases during mitosis. EMBO J 1992; 11(8):2895-2902.
89. Tsujimura K, Tanaka J, Ando S et al. Identification of phosphorylation sites on glial fibrillary acidic protein for cdc2 kinase and Ca(2+)-calmodulin-dependent protein kinase II. J Biochem (Tokyo) 1994; 116(2):426-434.
90. Kosako H, Amano M, Yanagida M et al. Phosphorylation of glial fibrillary acidic protein at the same sites by cleavage furrow kinase and Rho-associated kinase. J Biol Chem 1997; 272(16):10333-10336.
91. Nakamura Y, Takeda M, Nishimura T. Dynamics of bovine glial fibrillary acidic protein phosphorylation. Neurosci Lett 1996; 205(2):91-94.
92. Takemura M, Gomi H, Colucci-Guyon E et al. Protective role of phosphorylation in turnover of glial fibrillary acidic protein in mice. J Neurosci 2002; 22(16):6972-6979.
93. Yasui Y, Amano M, Nagata K et al. Roles of Rho-associated kinase in cytokinesis; mutations in Rho-associated kinase phosphorylation sites impair cytokinetic segregation of glial filaments. J Cell Biol 1998; 143(5):1249-1258.
94. Takemura M, Nishiyama H, Itohara S. Distribution of phosphorylated glial fibrillary acidic protein in the mouse central nervous system. Genes Cells 2002; 7(3):295-307.

95. Pekny M, Eliasson C, Chien CL et al. GFAP-deficient astrocytes are capable of stellation in vitro when cocultured with neurons and exhibit a reduced amount of intermediate filaments and an increased cell saturation density. Exp Cell Res 1998; 239(2):332-343.
96. Eliasson C, Sahlgren C, Berthold CH et al. Intermediate filament protein partnership in astrocytes. J Biol Chem 1999; 274(34):23996-24006.
97. Lepekhin EA, Eliasson C, Berthold CH et al. Intermediate filaments regulate astrocyte motility. J Neurochem 2001; 79(3):617-625.
98. Menet V, Gimenez y Ribotta M, Chauvet N et al. Inactivation of the glial fibrillary acidic protein gene, but not that of vimentin, improves neuronal survival and neurite growth by modifying adhesion molecule expression. J Neurosci 2001; 21(16):6147-6158.
99. Pekny M, Leveen P, Pekna M et al. Mice lacking glial fibrillary acidic protein display astrocytes devoid of intermediate filaments but develop and reproduce normally. EMBO J 1995; 14(8):1590-1598.
100. McCall MA, Gregg RG, Behringer RR et al. Targeted deletion in astrocyte intermediate filament (Gfap) alters neuronal physiology. Proc Natl Acad Sci USA 1996; 93(13):6361-6366.
101. Chou YH, Khuon S, Herrmann H et al. Nestin promotes the phosphorylation-dependent disassembly of vimentin intermediate filaments during mitosis. Mol Biol Cell 2003; 14(4):1468-1478.
102. Hirako Y, Yamakawa H, Tsujimura Y et al. Characterization of mammalian synemin, an intermediate filament protein present in all four classes of muscle cells and some neuroglial cells: Colocalization and interaction with type III intermediate filament proteins and keratins. Cell Tissue Res 2003; 313(2):195-207.
103. Hernandez MR, Agapova OA, Yang P et al. Differential gene expression in astrocytes from human normal and glaucomatous optic nerve head analyzed by cDNA microarray. Glia 2002; 38(1):45-64.
104. Eddleston M, Mucke L. Molecular profile of reactive astrocytes—implications for their role in neurologic disease. Neuroscience 1993; 54(1):15-36.
105. Pekny M. Astrocytic intermediate filaments: Lessons from GFAP and vimentin knock-out mice. Prog Brain Res 2001; 132:23-30.
106. Colucci-Guyon E, Portier MM, Dunia I et al. Mice lacking vimentin develop and reproduce without an obvious phenotype. Cell 1994; 79(4):679-694.
107. Gomi H, Yokoyama T, Fujimoto K et al. Mice devoid of the glial fibrillary acidic protein develop normally and are susceptible to scrapie prions. Neuron 1995; 14(1):29-41.
108. Tanaka H, Katoh A, Oguro K et al. Disturbance of hippocampal long-term potentiation after transient ischemia in GFAP deficient mice. J Neurosci Res 2002; 67(1):11-20.
109. Shibuki K, Gomi H, Chen L et al. Deficient cerebellar long-term depression, impaired eyeblink conditioning, and normal motor coordination in GFAP mutant mice. Neuron 1996; 16(3):587-599.
110. Colucci-Guyon E, Gimenez y Ribotta M, Maurice T et al. Cerebellar defect and impaired motor coordination in mice lacking vimentin. Glia 1999; 25:33–43.
111. Liedtke W, Edelmann W, Bieri PL et al. GFAP is necessary for the integrity of CNS white matter architecture and long-term maintenance of myelination. Neuron 1996; 4:607–615.
112. Liedtke W, Edelmann W, Chiu FC et al. Experimental autoimmune encephalomyelitis in mice lacking glial fibrillary acidic protein is characterized by a more severe clinical course and an infiltrative central nervous system lesion. Am J Pathol 1998; 152:251–259.
113. Pekny M, Johansson CB, Eliasson C et al. Abnormal reaction to central nervous system injury in mice lacking glial fibrillary acidic protein and vimentin. J Cell Biol 1999; 145(3):503-514.
114. Bush TG, Puvanachandra N, Horner CH et al. Leukocyte infiltration, neuronal degeneration, and neurite outgrowth after ablation of scar-forming, reactive astrocytes in adult transgenic mice. Neuron 1999; 2:297–308.
115. Faulkner JR, Herrmann JE, Woo MJ et al. Reactive astrocytes protect tissue and preserve function after spinal cord injury. J Neurosci 2004; 24:2143–2155.
116. Menet V, Prieto M, Privat A et al. Axonal plasticity and functional recovery after spinal cord injury in mice deficient in both glial fibrillary acidic protein and vimentin genes. Proc Natl Acad Sci USA 2003; 100:8999-9004.
117. Xu K, Malouf AT, Messing A et al. Glial fibrillary acidic protein is necessary for mature astrocytes to react to beta-amyloid. Glia 1999; 25:390-403.
118. Menet V, Gimenez YRM, Sandillon F et al. GFAP null astrocytes are a favorable substrate for neuronal survival and neurite growth. Glia 2000; 31:267–272.
119. Wang X, Messing A, David S. Axonal and nonneuronal cell responses to spinal cord injury in mice lacking glial fibrillary acidic protein. Exp Neurol 1997; 148:568-576.
120. Wilhelmsson, Li L, Pekna M et al. Absence of glial fibrillary acidic protein and vimentin prevents hypertrophy of astrocytic processes and improves post-traumatic regeneration. J Neurosci 2004; 24:5016-5021.

121. Ishikawa N, Takemura M, Koyama Y et al. Endothelins promote the activation of astrocytes in rat neostriatum through ET(B) receptors. Eur J Neurosci 1997; 9:895–901.
122. Baba A. Role of endothelin B receptor signals in reactive astrocytes. Life Sci 1998; 62:1711–1715.
123. Koyama Y, Takemura M, Fujiki K et al. BQ788, an endothelin ET(B) receptor antagonist, attenuates stab wound injury-induced reactive astrocytes in rat brain. Glia 1999; 26:268–271.
124. Peters CM, Rogers SD, Pomonis JD et al. Endothelin receptor expression in the normal and injured spinal cord: Potential involvement in injury-induced ischemia and gliosis. Exp Neurol 2003; 180:1-13.
125. Johansson CB, Momma S, Clarke DL et al. Identification of a neural stem cell in the adult mammalian central nervous system. Cell 1999; 96(1):25-34.
126. Colucci-Guyon E, Portier MM, Dunia I et al. Mice lacking vimentin develop and reproduce without an obvious phenotype. Cell 1994; 79:679–694.
127. Eckes B, Dogic D, Colucci-Guyon E et al. Impaired mechanical stability, migration and contractile capacity in vimentin-deficient fibroblasts. J Cell Sci 1998; 111(Pt 13):1897-1907.
128. Holwell TA, Schweitzer SC, Evans RM. Tetracycline regulated expression of vimentin in fibroblasts derived from vimentin null mice. J Cell Sci 1997; 110:1947–1956.
129. Okabe M, Ikawa M, Kominami K et al. 'Green mice' as a source of ubiquitous green cells. FEBS Lett 1997; 407:313–319.
130. Kinouchi R, Takeda M, Yang L et al. Robust neural integration from retinal transplants in mice deficient in GFAP and vimentin. Nat Neurosci 2003; 6(8):863-868.
131. Pekny M, Pekna M, Wilhelmsson U et al. Response to quinlan and nilsson: Astroglia sitting at the controls? Trends Neurosci 2004; 27:243-244.
132. Quinlan R, Nilsson M. Reloading the retina by modifying the glial matrix. Trends Neurosci 2004; 27:241-242.
133. Emsley JG, Arlotta P, Macklis JD. Star-cross'd neurons: Astroglial effects on neural repair in the adult mammalian CNS. Trends Neurosci 2004; 27:238–240.
134. Fuchs E, Cleveland DW. A structural scaffolding of intermediate filaments in health and disease. Science 1998; 279:514–519.
135. Lane B, Pekny M. Stress models for the study of intermediate filament function. In: Omary MB, Coulombe PA, eds. Intermediate Filament Cytoskeleton 2004, in press.
136. Nawashiro H, Messing A, Azzam N et al. Mice lacking GFAP are hypersensitive to traumatic cerebrospinal injury. Neuroreport 1998; 9:1691–1696.
137. Lundkvist A, Reichenbach A, Betsholtz C et al. Under stress, the absence of intermediate filaments from Muller cells in the retina has structural and functional consequences. J Cell Sci 2004, in press.
138. Hoffman E. Volume regulation in cultured cells. New York: Academic Press, 1991.
139. Kimelberg HK. Swelling and volume control in brain astroglial cells. New York: Springer, 1991.
140. Pasantes-Morales H, Moran J, Schousboe A. Volume-sensitive release of taurine from cultured astrocytes: Properties and mechanism. Glia 1990; 3:427–432.
141. Vitarella D, DiRisio DJ, Kimelberg HK et al. Potassium and taurine release are highly correlated with regulatory volume decrease in neonatal primary rat astrocyte cultures. J Neurochem 1994; 63(3):1143-1149.
142. Moran J, Maar T, Pasantes-Morales H. Cell volume regulation in taurine deficient cultured astrocytes. Adv Exp Med Biol 1994; 359:361-367.
143. Sanchez-Olea R, Moran J, Schousboe A et al. Hyposmolarity-activated fluxes of taurine in astrocytes are mediated by diffusion. Neurosci Lett 1991; 130(2):233-236.
144. Cantiello HF, Prat AG, Bonventre JV et al. Actin-binding protein contributes to cell volume regulatory ion channel activation in melanoma cells. J Biol Chem 1993; 268:4596–4599.
145. Cantiello HF. Role of actin filament organization in cell volume and ion channel regulation. J Exp Zool 1997; 279(5):425–435.
146. Moran J, Sabanero M, Meza I et al. Changes of actin cytoskeleton during swelling and regulatory volume decrease in cultured astrocytes. Am J Physiol 1996; 271(6 Pt 1):C1901-1907.
147. Ding M, Eliasson C, Betsholtz C et al. Altered taurine release following hypotonic stress in astrocytes from mice deficient for GFAP and vimentin. Brain Res Mol Brain Res 1998; 62:7-81.
148. Anderova M, Kubinova S, Mazel T et al. Effect of elevated K(+), hypotonic stress, and cortical spreading depression on astrocyte swelling in GFAP-deficient mice. Glia 2001; 35(3):189-203.
149. Pekny M, Eliasson C, Siushansian R et al. The impact of genetic removal of GFAP and/or vimentin on glutamine levels and transport of glucose and ascorbate in astrocytes. Neurochem Res 1999; 24(11):1357-1362.

150. Hughes EG, Maguire JL, McMinn MT et al. Loss of glial fibrillary acidic protein results in decreased glutamate transport and inhibition of PKA-induced EAAT2 cell surface trafficking. Brain Res Mol Brain Res 2004; 124(2):114-123.
151. Smith LE, Wesolowski E, McLellan A et al. Oxygen-induced retinopathy in the mouse. Invest Ophthalmol Vis Sci 1994; 35(1):101-111.
152. Bigner SH, McLendon RE, Al-dosari N et al. In: KW VBaK, ed. The Genetic Basis of Human Cancer. New York: McGraw-Hill, 1998:661-670.
153. Jacque CM, Kujas M, Poreau A et al. GFA and S 100 protein levels as an index for malignancy in human gliomas and neurinomas. J Natl Cancer Inst 1979; 62(3):479-483.
154. van der Meulen JD, Houthoff HJ, Ebels EJ. Glial fibrillary acidic protein in human gliomas. Neuropathol Appl Neurobiol 1978; 4(3):177-190.
155. Tascos NA, Parr J, Gonatas NK. Immunocytochemical study of the glial fibrillary acidic protein in human neoplasms of the central nervous system. Hum Pathol 1982; 13(5):454-458.
156. Velasco ME, Dahl D, Roessmann U et al. Immunohistochemical localization of glial fibrillary acidic protein in human glial neoplasms. Cancer 1980; 45(3):484-494.
157. Jacque CM, Vinner C, Kujas M et al. Determination of glial fibrillary acidic protein (GFAP) in human brain tumors. J Neurol Sci 1978; 35(1):147-155.
158. Kajiwara K, Orita T, Nishizaki T et al. Glial fibrillary acidic protein (GFAP) expression and nucleolar organizer regions (NORs) in human gliomas. Brain Res 1992; 572(1-2):314-318.
159. Hara A, Sakai N, Yamada H et al. Proliferative assessment of GFAP-positive and GFAP-negative glioma cells by nucleolar organizer region staining. Surg Neurol 1991; 36(3):190-194.
160. Rutka JT, Hubbard SL, Fukuyama K et al. Effects of antisense glial fibrillary acidic protein complementary DNA on the growth, invasion, and adhesion of human astrocytoma cells. Cancer Res 1994; 54(12):3267-3272.
161. Weinstein DE, Shelanski ML, Liem RK. Suppression by antisense mRNA demonstrates a requirement for the glial fibrillary acidic protein in the formation of stable astrocytic processes in response to neurons. J Cell Biol 1991; 112(6):1205-1213.
162. Chen WJ, Liem RK. Reexpression of glial fibrillary acidic protein rescues the ability of astrocytoma cells to form processes in response to neurons. J Cell Biol 1994; 127(3):813-823.
163. Rutka JT, Smith SL. Transfection of human astrocytoma cells with glial fibrillary acidic protein complementary DNA: Analysis of expression, proliferation, and tumorigenicity. Cancer Res 1993; 53(15):3624-3631.
164. Toda M, Miura M, Asou H et al. Cell growth suppression of astrocytoma C6 cells by glial fibrillary acidic protein cDNA transfection. J Neurochem 1994; 63(5):1975-1978.
165. Leonard JR, D'Sa C, Klocke BJ et al. Neural precursor cell apoptosis and glial tumorigenesis following transplacental ethyl-nitrosourea exposure. Oncogene 2001; 20(57):8281-8286.
166. Oda H, Zhang S, Tsurutani N et al. Loss of p53 is an early event in induction of brain tumors in mice by transplacental carcinogen exposure. Cancer Res 1997; 57(4):646-650.
167. Wilhelmsson U, Eliasson C, Bjerkvig R et al. Loss of GFAP expression in high-grade astrocytomas does not contribute to tumor development or progression. Oncogene 2003; 22(22):3407-3411.
168. Messing A, Head MW, Galles K et al. Fatal encephalopathy with astrocyte inclusions in GFAP transgenic mice. Am J Pathol 1998; 152:391-398.
169. Brenner M, Johnson AB, Boespflug-Tanguy O et al. Mutations in GFAP, encoding glial fibrillary acidic protein, are associated with Alexander disease. Nat Genet 2001; 27:117-120.
170. Li R, Messing A, Goldman JE et al. GFAP mutations in Alexander disease. Int J Dev Neurosci 2002; 20:259-268.
171. Okamoto Y, Mitsuyama H, Jonosono M et al. Autosomal dominant palatal myoclonus and spinal cord atrophy. J Neurol Sci 2002; 195:71-76.
172. Namekawa M, Takiyama Y, Aoki Y et al. Identification of GFAP gene mutation in hereditary adult-onset Alexander's disease. Ann Neurol 2002; 52:779-785.
173. Messing A, Brenner M. GFAP: Functional implications gleaned from studies of genetically engineered mice. Glia 2003; 43:87-90.
174. Messing A, Brenner M. Alexander disease: GFAP mutations unify young and old. Lancet Neurol 2003; 2:75.
175. Wilhelmsson U. Astrocytes, reactive gliosis and CNS regeneration [Dr. Med. Sc. thesis]. Sweden: Sahlgrenska Academy at Göteborg University, 2004.
176. Enge M, Wilhelmsson U, Abramsson A et al. Neuron-specific ablation of PDGF-B is compatible with normal central nervous system development and astroglial response to injury. Neurochem Res 2003; 28(2):271-279.
177. Sternberg SS. Histology for Pathologists. 2nd ed. Lippincott: Williams and Wilkins, 1997.

CHAPTER 3

Neuronal Intermediate Filaments and Neurodegenerative Diseases

Gee Y. Ching and Ronald K.H. Liem*

Abstract

Neuronal intermediate filaments (IFs) are major components of the cytoskeleton in neurons and are composed of a family of neuronal IF proteins that include peripherin, α-internexin, and neurofilament triplet proteins (NFTPs), designated NF-L, NF-M, and NF-H for low, medium, and high molecular weight subunits, respectively. NFTPs are determinants of axonal caliber, while α-internexin may play a role in neuronal regeneration and peripherin in proper development of a defined population of sensory neurons. The composition and proper stoichiometry of these proteins are important for IF assembly and transport. Abnormal IF accumulations in neuronal perikarya and proximal axons are pathological hallmarks of many neurodegenerative diseases. Studies of transgenic mouse models as well as identification of NF-H mutations in sporadic amyotrophic lateral sclerosis (ALS) and NF-L mutations in Charcot-Marie-Tooth disease indicate that neuronal IFs have a direct role in pathogenesis of these diseases. Overexpression of individual NFTP subunit or peripherin in transgenic mice causes abnormal accumulation of IFs in motor neurons and motor dysfunction, reminiscent of ALS, whereas overexpression of α-internexin causes Purkinje cell death and motor coordination deficits. Deletion of NF-L or overexpression of NF-H extends the lifespan of transgenic mice expressing mutated Cu/Zn superoxide dismutase (SOD1), suggesting that NFTPs may also contribute to SOD1 mutation-linked ALS. The misaccumulated IFs in neuronal perikarya and proximal axons are disorganized and aberrantly phosphorylated. The kinases and phosphatases that are known to control phosphorylation of NFTPs are deregulated in the disease state. Studies of transgenic mouse models show that axonal transport of neuronal IFs and other cellular components is impaired in the disease state, ultimately leading to neurodegeneration. Formation of disorganized and aberrantly phospohorylated IFs, changes in stoichiometric levels of neuronal IF proteins due to dysregulation of their gene expression, and reduced levels of functional motor proteins may contribute in part to this defective axonal transport. Recent studies raise the possibility that cis-acting and trans-acting determinants of NF-L mRNA stability may also be involved in degeneration of motor neurons. Taken together, these data suggest that abnormal IF accumulations in neurons may not simply be the by-products of neurodegenerative diseases but may instead play a contributory role in pathogenesis of these diseases.

*Corresponding Author: Ronald K.H. Liem— Department of Pathology, Columbia University College of Physicians and Surgeons, 630 West 168th Street, New York, New York 10032, U.S.A. Email: rkl2@columbia.edu

Intermediate Filaments, edited by Jesus Paramio. ©2006 Landes Bioscience and Springer Science+Business Media.

Neuronal Intermediate Filament Proteins

Structure and Assembly

Neuronal intermediate filament (IF) proteins form the 10 nm filaments present in the cytoskeleton of neurons. In the mammalian nervous system seven IF proteins have been identified and like other members of the IF protein family, they can be classified into distinct IF types according to sequence homology and the exon-intron organization of their genes. α-Internexin and the neurofilament triplet proteins (NFTPs), designated as NF-L, NF-M and NF-H for low, medium, and high molecular weight subunits, respectively, are type IV; peripherin and vimentin are type III; and nestin is type VI. These proteins share a tripartite structure typical of all IF proteins, which consists of a highly conserved α-helical rod domain flanked by variable amino-terminal head and carboxyl-terminal tail domains (Fig. 1). The α-helical rod domain contains hydrophobic heptad repeats believed to mediate the formation of coiled-coil dimers, which associate laterally and longitudinally to form protofilaments (tetramers), protofibrils (octamers) and ultimately 10 nm IFs (for reviews, see refs 1-4).[1-4]

In mature mammalian neurons, the IFs are composed of α-internexin, NFTPs, and peripherin. Assembly studies performed with transfected cells demonstrate that α-internexin and peripherin can self-assemble into IFs and coassemble with NFTPs.[5-9] In contrast, the NFTPs

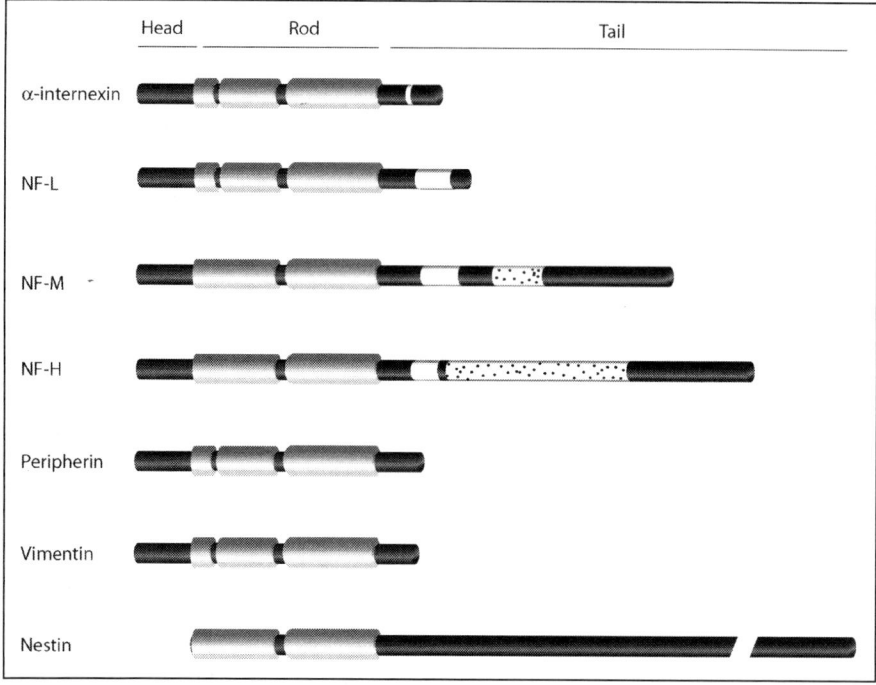

Figure 1. Schematic illustration of the tripartite structure of neuronal IF proteins. The highly conserved central α-helical rod domain (grey) is flanked by variable N-terminal head and C-terminal tail domains (black). Coils 1A, 1B and 2 within the rod domain are separated by small linker regions (black). NF-M, NF-H and nestin contain coils 1 and 2. A glutamic acid-rich region (white) is present in the tail domains of α-internexin and NFTPs. A region (dotted area) consisting of up to 13 and 50 KSP repeats, respectively, is present in the tail domains of NF-M and NF-H. The tail domain of nestin is too long to be illustrated within the scale shown here.

are obligate heteropolymers since NF-M or NF-H requires NF-L for coassembly into IFs.[5,10] Human NF-L can partially self-assemble into filaments, whereas rodent NF-L cannot.[5,10-12] The head and tail domains of α-internexin and NFTPs may contribute to the differences in assembly properties between these proteins.[6,13]

Expression

The neuronal IF proteins are expressed differentially during neurogenesis. Vimentin and nestin are expressed only in neuroepithelial stem cells,[14] whereas α-internexin, NFTPs and peripherin are expressed in post-mitotic neurons. α-Internexin is expressed earlier and more abundantly than NFTPs in the developing peripheral (PNS) and central (CNS) nervous systems.[15-17] As development continues, the levels of α-internexin decrease whereas those of the NFTPs increase. By adulthood, α-internexin is found primarily in the CNS and very minimally in the PNS, while the NFTPs are present abundantly in both the CNS and PNS. In adult CNS, α-internexin is colocalized with NFTPs in most axons but is present at lower levels in large neurons, such as the motor neurons of the spinal cord and cranial nerve ganglia. However, it is highly expressed in the cerebellar granule cells, which lack NFTPs.[16,18] Peripherin is expressed predominantly in the PNS but is also found at low levels in subsets of CNS neurons, such as the motor neurons of the spinal cord.[19-22]

Function

The conduction velocity of an impulse down the axon is determined, in part, by the axonal caliber. Fast impulse conduction velocity is crucial for the proper functioning of axons with large diameter, such as motor axons. NFTPs are present at high levels in large axons. The carboxyl-terminal tail domains of NF-M and NF-H, which form the side arms projected laterally from the filament core, contain up to 13 and 50 lysine-serine-proline (KSP) repeats, respectively, depending on the species. Phosphorylation of these KSP repeats is believed to regulate interfilament spacing and thereby axonal caliber.[23,24]

The role of NFTPs in regulating axonal caliber has been investigated in mice with targeted gene disruption. Surprisingly, targeted disruption of the NF-H gene results in little or modest effects on axonal caliber.[25-27] Furthermore, targeted deletion of the NF-H tail domain and all of its phosphorylation sites also does not affect the axonal diameter of motor and sensory neurons although acquisition of normal axonal caliber is slower in the mice.[28] In contrast, the absence of NF-L or NF-M, which have fewer phosphorylation sites in their tail domains, has more impact on the radial growth of axons. Mice with targeted disruption of NF-L,[29] NF-M[30] or both NF-M and NF-H genes[31] show a dramatic decrease in IF content and a 2-3 fold reduction in axonal caliber. Interestingly, the absence of NF-L causes a 20-fold reduction in the levels of NF-M and NF-H proteins,[29] and the absence of NF-M results in a substantial decrease in the levels of NF-L but an increase in the levels of NF-H.[30] The lack of both NF-M and NF-H causes sequestering of unassembled NF-L in neuronal perikarya.[31] Overall, these data suggest that the composition and proper stoichiometry of NFTPs are important for normal IF assembly and transport as well as axonal caliber. Moreover, phosphorylation of the NF-M tail domain may be more important than that of NF-H in determining axonal caliber. This conclusion is further supported by analysis of transgenic mice overexpressing individual NFTP or a combination of NFTP subunits.[32] Overexpressing each NFTP subunit individually causes inhibition of radial growth of axons, whereas overexpressing NF-L together with either NF-M or NF-H leads to an increase. Recent studies with NF-M tail domain-deleted mice, which contain normal ratios of the three NFTP subunits, suggest that the NF-M tail domain may regulate axonal caliber by forming long cross-bridges extended laterally from the filament core and thereby increasing interfilament spacing.[33]

Targeted disruption of the α-internexin[34] or peripherin genes[35] have no effect on axonal caliber. Instead, the α-internexin null mice develop normally with no obvious changes in the nervous system. However, a recent report of up-regulation of the α-internexin expression in injured motor neurons suggest that this protein may play a role in neuronal regeneration.[36]

The peripherin null mice display a reduced number of unmyelinated sensory axons, indicating its role in proper development of a defined population of sensory neurons.

Overall, none of the null mice described above exhibit gross developmental defects and overt behavioral phenotypes, demonstrating that these neuronal IF proteins are not essential for the development of gross structure of the nervous system and survival of these animals. Nevertheless, deficiencies in these proteins have deleterious effects on the null mice. The NF-L null mice exhibit a 20% loss of motor neurons and delayed maturation of regenerating myelinated axons.[29] The mice with null mutation in NF-M develop atrophy of motor axons and a hind limb paralysis with aging.[37] Increase in microtubule content[27,30,31] and alteration in slow axonal transport[27,31] are observed in the mice with null mutations in either NF-M, NF-H, or both NF-M and NF-H.

Neuronal IFs in Neurodegenerative Diseases

Abnormal IF Accumulations in Neurons

Abnormal accumulations of neuronal IFs are pathological hallmarks of many neurodegenerative diseases, such as amyotrophic lateral sclerosis (ALS), Lewy body-type dementias, and Parkinson's disease.[38,39] High levels of neuronal IFs are accumulated in the perikarya and axons of motor neurons of ALS patients. NFTPs and peripherin are detected in these neuronal IF inclusions (Fig. 2).[40,41]

The relationship between neuronal IFs and neuronal degeneration has been extensively studied in transgenic mouse models. Transgenic mice that overexpress NFTP subunits individually exhibit massive accumulations of neuronal IFs in motor neuron perikarya and in proximal portions of motor axons.[42-48] In some cases, such as overexpression of mutated mouse NF-L, wild-type human NF-M or NF-H, these motor neuron pathologies are accompanied by skeletal muscle atrophy and motor dysfunction, reminiscent of those found in human ALS.[42,45,48] Despite these characteristics, most of the transgenic mice do not display significant loss of motor neurons, an important difference with respect to the human disease. Although overexpression of the mutated mouse NF-L results in motor neuron death,[45]

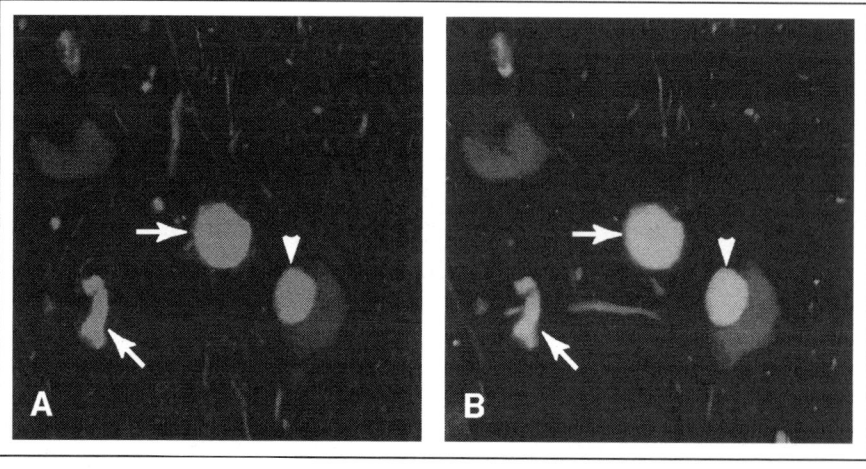

Figure 2. Abnormal accumulation of neuronal IFs in spinal motor neurons of a sporadic ALS patient. Immunofluorescence stainings with antibodies to (A) NF-H or (B) peripherin reveal abnormal IF accumulations in the cell body (arrowhead) and axonal spheroids (arrows). (These photographs are courtesy of Dr. Arthur Hayes, Department of Pathology, Columbia University College of Physicians and Surgeons.)

overexpression of wild-type mouse NF-L does not.[43] Some of the muscle atrophy observed in the mice overexpressing the wild-type NF-L could be attributed to the experimental use of a viral promoter that expresses NF-L ubiquitously.[43] It is noteworthy that overexpression of wild-type human NF-H[42] or human NF-M[48] induces motor dysfunction and in the case of human NF-M, loss of motor neurons as well, whereas overexpression of wild-type mouse NF-H[47] or NF-M[46] does not. These differences may be partly due to variation between the two species in the sequence and number of multiple KSP repeats in the carboxyl tail domains of NF-M and NF-H.[49,50]

Transgenic mice overexpressing peripherin exhibit IF inclusions reminiscent of axonal spheroids detected in ALS patients as well as motor neuron death during aging.[51] In contrast to NFTPs and peripherin, transgenic mice overexpressing α-internexin do not display ALS-like phenotype.[52] The lack of effects on motor neurons may be due to intrinsically low levels of α-internexin expression in these neurons.[16,18] Instead, overexpression of α-internexin causes IF accumulations in perikarya and axons of Purkinje cells, leading to the progressive neurodegeneration and ultimate loss of the neurons. The transgenic mice have motor coordination deficits preceding neuron loss, a pattern characteristic of some human neurodegenerative diseases. The degrees of motor coordination deficits and neuron loss are transgene dosage-dependent, demonstrating a direct role of levels of misaccumulated neuronal IFs in neurodegeneration.

Studies of various transgenic mouse models suggest that direct mutations or dysregulated expression of neuronal IF genes, abnormal phosphorylation of neuronal IF proteins, defects in axonal transport and in protein degradation can lead to accumulation of IF inclusions. Misaccumulation of neuronal IFs, in turn, can affect neuronal IF organization, phosphorylation, and axonal transport, ultimately leading to neuronal dysfunction. Thus, abnormal IF accumulations in neurons may not simply be the by-products of neurodegenerative diseases but may instead play a contributory role in pathogenesis of these diseases, especially in light of the identification of mutations in the NFTP genes in some neurodegenerative diseases.

Neuronal IF Gene Mutations in Human Diseases

ALS is an adult-onset neurological disease pathologically characterized by a loss of motor neurons in the cerebral cortex and the spinal cord, leading to skeletal muscle atrophy, paralysis, and ultimately death within 2-5 years after the onset of the disease. Approximately 10% of ALS cases are familial, with the remainder being sporadic. Mutations in the Cu/Zn superoxide dismutase gene (SOD1) have been identified that account for 20% of the familial ALS cases.[53] Transgenic mice with ALS-linked mutations in SOD1 exhibit abnormal IF accumulations and degeneration of motor neurons.[54] While the causes of sporadic ALS remain largely unknown, mutations in the NF-H gene have been identified in a small number of sporadic ALS patients. All of these mutations are deletions or insertions in the carboxyl tail domain of NF-H (Fig. 3A).[55-57]

Charcot-Marie-Tooth disease (CMT) is a hereditary peripheral neuropathy affecting both motor and sensory neurons. CMT has been classified into types 1 (CMT1), 2 (CMT2), 3 (CMT3), and 4 (CMT4): types 1, 3, and 4 are myelinopathies and type 2 is an axonopathy.[58] Mutations in the NF-L gene have recently been found in some cases of CMT2E, a dominantly inherited neuropathy, and all but one of them are mis-sense mutations (Fig. 3B).[59-63]

Parkinson's disease is a progressive neurological disorder that affects the dopaminergic neurons in the substantia nigra, with clinical symptoms such as resting tremor and rigid movement. Mutations in the Parkin gene and the alpha-synuclein gene have been reported to be responsible for the disease.[64] Recently, three variants in the NF-M gene are found in three patients with Parkinson's disease[65,66] However, the involvement of mutations in NF-M in this human disease remains to be determined.

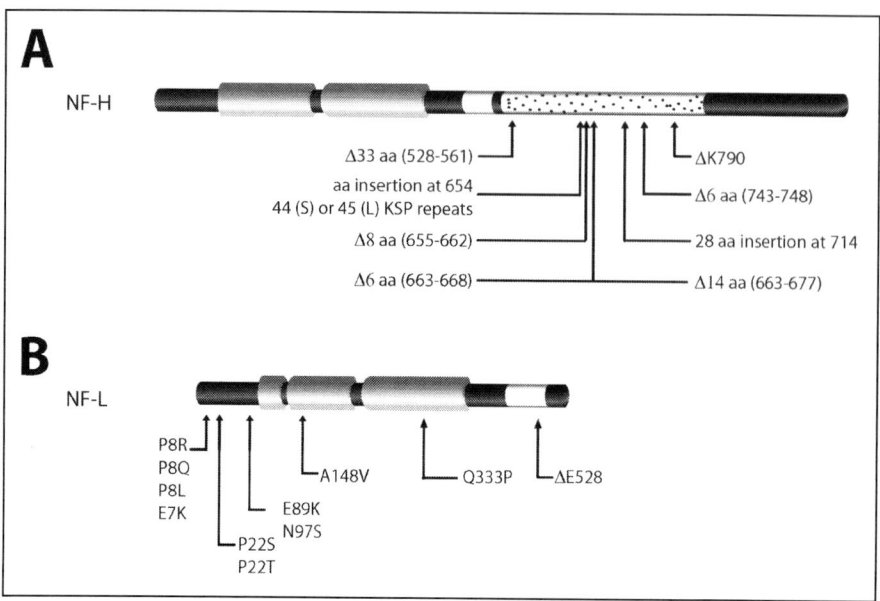

Figure 3. Schematic illustration of the neuronal IF protein mutations found in human neurodegenerative diseases. A) Deletion (Δ) and insertion mutations of NF-H are found in a small number of patients with sporadic ALS and they are located in the protein region containing multiple KSP repeats. B) Mutations in NF-L are found in patients with CMT2E. With the exception of one deletion (Δ), all the CMT2E mutations are missense mutations.

Dysregulation of Neuronal IF Gene Expression

The roles of dysregulation of neuronal IF gene expression in neuronal degeneration are poorly understood. Alterations of the NF-L or peripherin mRNA levels in motor neurons of ALS patients have been reported.[67-70] Recent studies raise the possibility that neuronal IF mRNAs may play contributory roles in the pathogenesis of motor neuron degeneration.

Cis-acting destabilizing elements have been identified in the 3' untranslated regions (3'UTR) of the NF-L and NF-H mRNAs and similar RNA-binding protein complexes assemble on the two regions.[71-73] Deletion or disruption of these elements results in stabilization of the NF-L and NF-H mRNAs. Transgenic mice containing a 36 bp c-myc insert within the destabilizing element of the NF-L transgene exhibit extensive vacuolar changes in motor neurons.[74] Furthermore, transgenic mice with a chimeric transgene containing an enhanced green fluorescent protein (EGFP) reporter gene and the 3' UTR of NF-L mRNA show growth retardation, motor nerve fiber atrophy and motor impairment.[75] When the chimeric transgene is modified to have a 36 bp c-myc insert disrupting the destabilizing element of the NF-L 3'UTR, a similar but more severe phenotype is observed. In contrast, such a phenotype is not observed in mice carrying an EGFP transgene. Additional studies with cultured motor neurons show that expression of a chimeric transgene containing the NF-L 3'UTR fused to the EGFP reporter gene results in early degeneration of neuritic processes and accumulation of ubiquitinated aggregates in the perikarya of degenerating motor neurons.[76] Taken together, these data suggest that the cis-acting destabilizing element of NF-L mRNA may be involved in degeneration of motor neurons, but the possibility of toxic effects of prolonged high-level expression of EGFP due to the presence of the NF-L 3'UTR cannot be entirely excluded.

Using a 68-nucleotide RNA probe containing the NF-L cis-acting destabilizing element, a neuron-enriched guanine exchange factor (GEF), p190RhoGEF, has been isolated.[77] p190RhoGEF is up-regulated during postnatal development and expressed abundantly in large differentiated neurons, including motor neurons of the spinal cord.[78] It binds not only to the NF-L destabilizing element, but also to BC1, an untranslated 152-nucleotide polymerase III transcript that is thought to serve as a molecular scaffold for regulating transport and translation of neuronal mRNAs at dendritic sites. It also interacts with microtubules, c-Jun N-terminal Kinase-interactive protein-1 (JIP-1) and 14-3-3, and is proposed to be involved in regulation of the neuronal cytoskeleton. The identification of p190RhoGEF as the trans-acting factor binding to the NF-L destabilizing element is interesting, in light of the finding of mutation in the *ALS2* gene, which encodes a GEF, as the genetic cause of a recessively inherited ALS.[79-81] A recent report suggests that trans-acting factors, which bind to the NF-L 3'UTR and destabilize NF-L mRNA, are present in normal spinal cord but absent in ALS spinal cord homogenates, resulting in increased NF-L mRNA levels in ALS tissues.[82] Identification of these trans-acting factors and their roles in pathogenesis of ALS are yet to be established.

An alternatively spliced isoform of mouse peripherin (61 kDA), which contains a 32-amino acid insertion within the α-helical rod domain due to retention of intron-4, is proposed to contribute to degeneration of motor neurons in a mouse model of ALS.[70] An antibody specifically against this 61 kDA isoform is able to immunostain axonal spheroids of motor neurons in two cases of familial ALS. However, the corresponding intron of the human peripherin gene differs in size and sequence from that of mouse peripherin gene, and if not spliced out, would result in a truncated protein. It is therefore unclear what protein the antibody recognizes in the tissues of the ALS patients. Further studies are needed to confirm the role of this 61 kDa peripherin isoform in pathogenesis of ALS. Nevertheless, analysis of peripherin isoforms could provide additional information regarding mechanisms of motor neuron degeneration, and RNA splicing has been demonstrated to play an important role in human diseases, including neurological disorders.[83-85]

Disorganization of Neuronal IFs

In normal neurons, neuronal IFs are dense in axons but are sparse in dendrites and perikarya. In axons, the filaments are regularly aligned and spaced, with well developed cross-bridges (formed by the NF-M and NF-H side arms) linking them to each other and to other cytoskeletal components, including microfilaments, microtubules, and their associated proteins. The IF structures are less organized in dendrites and even less so in perikarya.[23] The tail domains of NF-M and NF-H, which form the side arms projected laterally from the filament core, are largely nonphosphorylated in dendrites, perikarya, and proximal parts of axons, but become highly phosphorylated in distal regions of axons.[86-88]

In diseased neurons, which contain abnormal IF accumulations in their perikarya and proximal parts of axons, the misaccumulated IFs are disorganized and contain less-developed cross-bridges. Numerous cellular organelles (such as mitochondria and smooth endoplasmic reticulum) are sequestered or entrapped in these neuronal IF accumulations and some of them look abnormal[44,52,89] (Fig. 4). Many of the misaccumulated IFs contain highly phosphorylated NF-M and NF-H, whose presence in neuronal perikarya is a pathological hallmark of many neurodegenerative diseases.[89-91] Disorganization of neuronal IFs may affect their interaction with other proteins, cytoskeletal components and membranous organelles. Excessive accumulations of these disorganized neuronal IFs may cause defects in axonal transport of IFs and other cellular components crucial for neuronal function.[47,92-94] Organelle damages and aberrant distribution of normal organelles may perturb metabolic activities of the affected neurons. The defects in axonal transport and perturbation in cellular metabolism contribute in part to neuronal dysfunction and degeneration.

Ultrastructural studies reveal cross-bridges from mitochondria to neuronal IFs or microtubules.[95] In neurons, mitochondria are transported in association with microtubules, actin filaments, and IFs.[96-98] Docking of mitochondria to neuronal IFs and microtubules influences the motility and distribution of mitochondria in axons. In vitro studies have demonstrated the direct binding of mitochondria to NF-M and NF-H and phosphorylation of the NF-M and NF-H side arms affects these interactions.[98,99] Phosphorylated neurofilaments bind mitochondria at

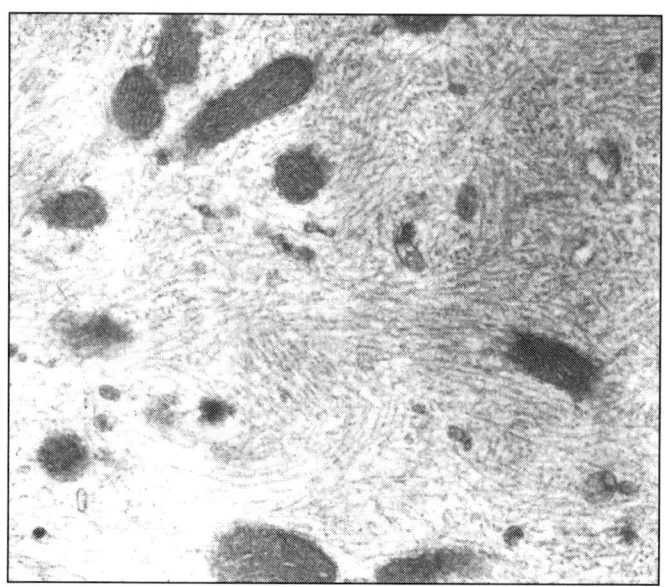

Figure 4. Electron micrograph of a swollen Purkinje cell axon from a transgenic mouse overexpressing rat α-internexin. Closely packed, randomly oriented neuronal IFs and sequestered organelles are observed in a cerebellar torpedo, which is an abnormal swelling of Purkinje cell axon.

high membrane potential and are released from them at low membrane potential, whereas de-phosphorylated neurofilaments bind mitochondria of low and high membrane potentials with similar efficiency.[98] Mitochondria are the principal source of energy for the cell and cytoskeletal changes that alter their distribution can have detrimental effects on the function and viability of cells. Motor neurons have high energy demands because of their large size and long axons. Therefore, they are particularly vulnerable to alteration in transport and distribution of mitochondria as well as to mitochondrial dysfunction that are induced by misaccumulation of disorganized and aberrantly phophorylated neuronal IFs.

Aberrant Phosphorylation of Neuronal IF Proteins

Neuronal IF proteins are continuously modified by various protein kinases and phosphatases that control phosphorylation at their amino-terminal head and carboxyl-terminal tail domains in axons. NF-M and particularly NF-H are highly phosphorylated due to the presence of multiple KSP repeats in their tail domains. These KSP repeats are phosphorylated by proline-directed kinases that include the cyclin-dependent Cdk5/p35, the extracellular signal-regulated Erk1/2, the glycogen synthase kinase-3 (GSK3) and the stress-activated protein kinases (SAPK).[100-103] The glutamic acid-rich region of the tail domain, shared by all NFTP subunits and α-internexin, also contains serine/threonine residues that can be phosphorylated by second messenger independent kinases such as casein kinases I and II (CKI/II).[104-106] Some of the serine/threonine residues in the head domains of NFTPs, α-internexin and peripherin are phosphorylated by second messenger dependent kinases such as Protein kinase-A and –C (PKA, PKC) and Ca^{2+}/ calmodulin-dependent protein kinase II (CAMKII).[107-113] Protein phosphatase-2A is implicated to dephosphorylate the KSP repeats in the tail domains of NF-H and NF-M, and protein phosphatase-1 is shown to dephosphorylate the head domain of NF-L.[114-116] Phosphorylation of NFTPs has significant roles in regulating axonal caliber[23,24,33] and cross-bridge formation,[117-119] IF assembly,[120] interaction with other cytoskeletal components,[121,122] axonal transport,[47,123] and protection of NFTPs from degradation by protease.[124,125]

Phosphorylation of the KSP repeats in the tail domains of NF-M and NF-H is regulated topographically during neuronal development. It is also correlated with myelination. NF-M and NF-H are synthesized in neuronal perikarya and their phosphorylation begins soon after synthesis. They are hypophosphorylated in the perikarya and become variably phosphorylated after they enter the axons. The two proteins are least phophorylated in the unmyelinated initial segment and the nodes of Ranvier where axonal caliber is smaller, but are heavily phosphorylated in the internode region of myelinated axons.[86-88,126-129] The detection of highly phosphorylated NF-M and NF-H in the perikaryal IF accumulations of the diseased neurons, as seen in neurodegenerative diseases and transgenic mouse models, suggest that the mechanisms regulating the phosphorylation of these proteins may be perturbed. Recent studies have provided evidence to link pathogenesis of neurodegenerative diseases with de-regulation of kinases and phosphatases that are known to modulate phosphorylation of neuronal IFs. Activation of the p38 mitogen-activated protein kinase (p38MAPK) is shown to occur in transgenic mice expressing a mutant SOD1 (G93A). Furthermore, p38MAPK colocalizes with phosphorylated NF-M and NF-H in vacuolized perikarya and neurites of spinal motor neurons.[130] Activation of p38MAPK is also detected in Alzheimer's and other tau-related diseases.[131] Hyperphosphorylation of NF-M and NF-H is associated with de-regulation of Cdk5 in transgenic mice expressing a mutant SOD1 (G37R).[132] Abnormal subcellular localization of Cdk5 is also observed in some neurodegenerative diseases, including ALS and Alzheimer's disease.[133,134] Overexpression of p25, a truncated form of the Cdk5 activator p35, causes hyperphosphorylation of neurofilaments and tau in transgenic mice[135] and accumulation of p25 is detected in the brains of Alzheimer's disease.[136] Calcineurin, a Ca^{2+}/calmodulin-dependent phosphatase-2B known to dephosphorylate NFTPs, is protected from oxidative inactivation by wild-type SOD1, but its activity is impaired in the motor cortex of brains from mutant SOD1-expressing transgenic mice.[137] Myotubularin-related 2 protein phospohatase (MTMR2), an ubiquitously expressed phosphatase that has enzyme activity toward phospholipids, is shown to interact with NF-L in both neurons and Schwann cells.[138] MTMR2 expression is absent or limited in CMT4B patients with demyelinating peripheral neuropathy due to MTMR2 mutations. Even when it is detectable in the patients, it is restricted to Schwann cell cytoplasm, and it is not detectable in the axon. The interaction between NF-L and MTMR2 might explain why the CMT4B pathology is restricted to the nervous system despite the ubiquitous expression of MTMR2. Whether NF-L is a substrate to be dephosphorylated by this phosphatase and the relevance of their interaction in the CMT4B pathogenesis are yet to be determined.

Defective Axonal Transport

Axons, such as those of motor and sensory neurons, can extend in length of one meter or more. Axonal proteins, membranous organelles and Golgi-derived vesicles are transported from neuronal perikarya to nerve terminals along the axon via a process called axonal transport. Thus, axonal transport is essential for growth and maintenance of axons and proper functioning of neurons.

Axonal transport has been categorized as either fast or slow based on the relative rates at which various cargoes move. In axons, microtubules are unipolar, with their plus ends directed toward the synaptic terminals and minus ends toward the neuronal perikarya. Therefore, the direction of cargo movements along microtubules further distinguishes axonal transport as anterograde (from perikarya to synapse) and retrograde (from synapse to perikarya). Fast axonal transport involves transport of membranous organelles, mitochondria and vesicles of the secretory and endocytic pathways along microtubules at average rates of 50-400 mm day.$^{-1}$ Slow axonal transport is further divided into two components: slow component 'a' involves transport of proteins associated with neuronal IFs and microtubules (including neuronal IF proteins, tubulin, and tau proteins) at average rates of 0.3-3 mm day^{-1}; component 'b' involves transport of actin and cytosolic matrix proteins at average rates of 2-8 mm day^{-1}. Slow axonal transport has generally been considered to be anterograde, but recent studies with GFP-tagged neurofilaments suggest that the observed neurofilaments can move bidirectionally, although predominantly in an anterograde direction. Motor proteins of the dynein and kinesin super-families have been identified as responsible for axonal transport. Most dynein proteins move to the minus ends of microtubules, whereas most kinesin proteins move to the plus ends (for a review, see refs.139-142).

It has long been debated whether neuronal IF proteins move in the form of monomers, oligomers, or polymers. Recent studies show that efficient axonal transport in vivo mini-mally requires hetero-oligomer formation such as between NF-M and either NF-L or α-internexin.[143] These studies suggest a model, where slow axonal transport of neuronal IFs involves rapid movements interrupted by frequent long pauses, resulting in a net slow trans-port rate.[139,140] In this model, the axonal cytoskeleton is viewed as a stationary network in dynamic equilibrium with transported cytoskeletal proteins, and the stationary cytoskeleton may consist of cross-linked, relatively hyperphosphorylated neuronal IFs. Relatively hypophosphorylated neuronal IF proteins (in the form of monomers, oligomers, or poly-mers) that are part of the "moving wave" may then associate with and dissociate from the stationary cytoskeleton along the axon. Phosphorylation of neuronal IF proteins, particu-larly the NF-M and NF-H tail domains, promotes associations between neuronal IFs and dissociation of neuronal IFs from kinesin motors, resulting in slowing of axonal transport of neuronal IFs. In contrast to kinesin, dynein motors appear to preferentially associate with phosphorylated neuronal IFs.

Transgenic mice overexpressing neuronal IF proteins[42-48,51,52] or expressing human mu-tant SOD1[92,93,144-147] exhibit abnormal IF accumulations in neuronal perikarya and proxi-mal parts of axons. These findings suggest that axonal transport of neuronal IFs may be perturbed. Indeed, defective axonal transport is observed in some of these transgenic mice.[47,92-94,146,147] Several factors may account for the deficits in axonal transport. Accumu-lation of high levels of disorganized, randomly oriented neuronal IFs may over-burden the transport machinery, creating a "traffic jam" situation. Moreover, disorganization of the misaccumulated neuronal IFs may affect the interaction of neuronal IFs with microtubules or motor proteins, thereby disrupting axonal transport. Since phosphorylation of NF-M and NF-H retards axonal transport of neuronal IFs, the presence of aberrantly phosphorylated NF-M and NF-H in the abnormal IF accumulations further compounds the impairment of axonal transport. Transgenic mice expressing a mutant SOD1 (G93A) exhibit reduced levels of kinesin motor proteins in axons as well as an impairment in fast anterograde axonal trans-port preceding loss of spinal motor neurons.[147] This reduction in the levels of kinesin motor proteins in the disease state suggests that normal levels of functional motor proteins as well as their accessibility are important for axonal transport. Indeed, targeted disruption of KIF5A, a neuron-specific conventional kinesin heavy chain, causes a specific deficit in axonal trans-port of neuronal IFs, resulting in accumulations of neuronal IFs in the perikarya of periph-eral sensory neurons accompanied by reduction in sensory axon caliber in the null mice.[148] Since cytoskeletal alterations induced by misaccumulated neuronal IFs may affect motility, distribution, and function of mitochondria (described above), motor proteins could be dam-aged by oxidative stress. In the case of SOD1-mediated ALS, the toxic effects of mutated SOD1 could also cause damage to motor proteins. Alternatively, misaccumulated neuronal IFs can potentially act as a motor-protein sink to deplete motor proteins from transporting cargoes into axons.

Defective axonal transport seen in the disease state can also be directly caused by muta-tions in neuronal IF proteins or motor proteins. Insertion and deletion mutations in the tail domain of NF-H that account for a small number of sporadic ALS cases presumably affect the structure and phosphorylation of NF-H and could thereby impair axonal transport. Transfection studies show that the NF-L mutations associated with CMT2E disrupt IF as-sembly and cause formation of aggregates, leading to deficits in axonal transport.[12,149] A recent report links a mutation in the largest subunit (DCTN1) of dynactin, a multiprotein complex that associates with dynein, to a slowly progressive, autosomal dominant form of lower motor neuron disease without sensory symptoms.[150] Overexpression of dynamitin, the p50 subunit of dynactin, causes disassembly of the dynactin complex, neuronal IF accumulations, and inhibition (but not complete blockage) of retrograde axonal transport in spinal motor neurons of transgenic mice.[151] These transgenic mice develop a late-onset, progressive motor neuron disease. Mice with gene-targeted disruption of the cytoplasmic dynein heavy chain also display formation of Lewy-like inclusion bodies, motor neuron degeneration, and impairment in fast retrograde transport.[152] Since retrograde axonal transport of neuronal IFs is dynein-dependent,[139-141] accumulation of neuronal IFs detected in the dynamitin-overexpressing mice and in the dynein mutant mice may be partly attributable to the defect in retrograde transport.

Conclusion

Identification of mutations of NF-H in ALS and mutations of NF-L in CMT indicates a direct role of neuronal IFs in neurodegenerative diseases. Transgenic mouse models have provided useful experimental systems to investigate how alterations in neuronal IFs contribute to neuronal dysfunction and degeneration. Neuronal IF gene mutations can cause formation of disorganized IFs or aggregates as well as aberrant phosphorylation of neuronal IF proteins, resulting in impairment of axonal transport, which ultimately leads to neurodegeneration. Changes in the stoichiometric levels of neuronal IF proteins due to dysregulation of IF gene expression can also affect IF assembly and transport. Deletion of NF-L extends the lifespan of mutant SOD1(G85R) mice by 5-6 weeks and increasing NF-H expression extends the lifespan of mutant SOD1(G37R) mice by up to 6 months, indicating that NFTPs contribute to SOD1 mutation-linked ALS.[153,154] In the latter case, overexpression of human NF-H, which can potentially act as a phosphorylation sink for deregulated Cdk5 activity in the SOD1(G37R) mice, reduces levels of nuclear Cdk4 and retinoblastoma (Rb) protein phosphorylation, cell cycle regulators in the neuronal death pathway and alleviates disease.[155] Taken together, the data from the transgenic mouse models suggest that the mechanisms of neurodegeneration are complex and involve the interplay of many factors. Thus, elucidating these mechanisms remains to be a challenging area of scientific research.

With technological advances in molecular biology, identification of additional neuronal IF gene mutations associated with neurodegenerative diseases is expected and will undoubtedly provide a better understanding of their roles in the disease mechanisms. Regulation of neuronal IF gene expression remains poorly understood, although alterations in their expression levels have been reported in some human neurodegenerative diseases. To date, cis-acting elements that control the neuron-specific expression of these genes have not been determined and only a few elements that regulate their expression levels have been identified. The roles of NF-L mRNA stability and alternative spliced peripherin isoform in the pathogenesis of ALS also require further analysis.

Axonal transport of neuronal IFs and other cellular components are crucial for growth and maintenance of axons and proper functioning of neurons. Studies of transgenic mouse models show that axonal transport is impaired in the disease state and its impairment is evidently linked to pathogenesis of neurodegenerative diseases. Future studies should include identification of motor proteins responsible for transport of neuronal IFs and elucidation of the mechanisms that regulate their association and dissociation from these motors. Since phosphorylation of neuronal IF proteins plays an important role in this regulation, identification and characterization of the kinases and phosphatases that control their phosphorylation will help understanding how axonal transport becomes disrupted in the disease state.

Acknowledgements

We thank Mr. Daniel Cheng for his help in the preparation of this manuscript. This work is supported by grant NS15182 from the NIH.

References

1. Fliegner KH, Liem RK. Cellular and molecular biology of neuronal intermediate filaments. Int Rev Cytol 1991; 131:109-67.
2. Shaw G. Neurofilmanet proteins. In: Burgoyne RD, ed. The Neuronal Cytoskeleton. New York: Wiley-Liss, 1991:185-214.
3. Lee MK, Cleveland DW. Neuronal intermediate filaments. Annu Rev Neurosci 1996; 19:187-217.
4. Heins S, Aebi U. Making heads and tails of intermediate filament assembly, dynamics and networks. Curr Opin Cell Biol 1994; 6(1):25-33.
5. Ching GY, Liem RK. Assembly of type IV neuronal intermediate filaments in nonneuronal cells in the absence of preexisting cytoplasmic intermediate filaments. J Cell Biol 1993; 122(6):1323-35.
6. Ching G, Liem R. Roles of head and tail domains in (alpha)-internexin's self-assembly and coassembly with the neurofilament triplet proteins. J Cell Sci 1998; 111(Pt 3):321-33.
7. Cui C, Stambrook PJ, Parysek LM. Peripherin assembles into homopolymers in SW13 cells. J Cell Sci 1995; 108(Pt 10):3279-84.
8. Ho CL, Chin SS, Carnevale K et al. Translation initiation and assembly of peripherin in cultured cells. Eur J Cell Biol 1995; 68(2):103-12.
9. Beaulieu JM, Robertson J, Julien JP. Interactions between peripherin and neurofilaments in cultured cells: Disruption of peripherin assembly by the NF-M and NF-H subunits. Biochem Cell Biol 1999; 77(1):41-5.
10. Lee MK, Xu Z, Wong PC et al. Neurofilaments are obligate heteropolymers in vivo. J Cell Biol 1993; 122(6):1337-50.
11. Carter J, Gragerov A, Konvicka K et al. Neurofilament (NF) assembly; divergent characteristics of human and rodent NF-L subunits. J Biol Chem 1998; 273(9):5101-8.
12. Perez-Olle R, Leung CL, Liem RK. Effects of Charcot-Marie-tooth-linked mutations of the neurofilament light subunit on intermediate filament formation. J Cell Sci 2002; 115(Pt 24):4937-46.
13. Ching GY, Liem RK. Analysis of the roles of the head domains of type IV rat neuronal intermediate filament proteins in filament assembly using domain-swapped chimeric proteins. J Cell Sci 1999; 112(Pt 13):2233-40.
14. Lendahl U, Zimmerman LB, McKay RD. CNS stem cells express a new class of intermediate filament protein. Cell 1990; 60(4):585-95.
15. Chiu FC, Barnes EA, Das K et al. Characterization of a novel 66 kd subunit of mammalian neurofilaments. Neuron 1989; 2(5):1435-45.
16. Kaplan MP, Chin SS, Fliegner KH et al. Alpha-internexin, a novel neuronal intermediate filament protein, precedes the low molecular weight neurofilament protein (NF-L) in the developing rat brain. J Neurosci 1990; 10(8):2735-48.
17. Fliegner KH, Kaplan MP, Wood TL et al. Expression of the gene for the neuronal intermediate filament protein alpha-internexin coincides with the onset of neuronal differentiation in the developing rat nervous system. J Comp Neurol 1994; 342(2):161-73.
18. Chien CL, Liem RK. Characterization of the mouse gene encoding the neuronal intermediate filament protein alpha-internexin. Gene 1994; 149(2):289-92.
19. Portier MM, de Nechaud B, Gros F. Peripherin, a new member of the intermediate filament protein family. Dev Neurosci 1983; 6(6):335-44.
20. Parysek LM, McReynolds MA, Goldman RD et al. Some neural intermediate filaments contain both peripherin and the neurofilament proteins. J Neurosci Res 1991; 30(1):80-91.
21. Escurat M, Djabali K, Gumpel M et al. Differential expression of two neuronal intermediate-filament proteins, peripherin and the low-molecular-mass neurofilament protein (NF-L), during the development of the rat. J Neurosci 1990; 10(3):764-84.
22. Troy CM, Brown K, Greene LA et al. Ontogeny of the neuronal intermediate filament protein, peripherin, in the mouse embryo. Neuroscience 1990; 36(1):217-37.
23. Hirokawa N. Molecular architecture and dynamics of the neuronal cytoskeleton. In: Burgoyne RD, ed. The Neuronal Cytoskeleton. New York: Wiley-Liss, 1991:5-74.
24. Nixon RA, Sihag RK. Neurofilament phosphorylation: A new look at regulation and function. Trends Neurosci 1991; 14(11):501-6.
25. Elder GA, Friedrich VL, Kang C et al. Requirement of heavy neurofilament subunit in the development of axons with large calibers. J Cell Biol 1998; 143(1):195-205.
26. Rao MV, Houseweart MK, Williamson TL et al. Neurofilament-dependent radial growth of motor axons and axonal organization of neurofilaments does not require the neurofilament heavy subunit (NF-H) or its phosphorylation. J Cell Biol 1998; 143(1):171-81.
27. Zhu Q, Lindenbaum M, Levavasseur F et al. Disruption of the NF-H gene increases axonal microtubule content and velocity of neurofilament transport: Relief of axonopathy resulting from the toxin beta,beta'-iminodipropionitrile. J Cell Biol 1998; 143(1):183-93.

28. Rao MV, Garcia ML, Miyazaki Y et al. Gene replacement in mice reveals that the heavily phosphorylated tail of neurofilament heavy subunit does not affect axonal caliber or the transit of cargoes in slow axonal transport. J Cell Biol 2002; 158(4):681-93.
29. Zhu Q, Coouillard-Despres S, Julien J-P. Delayed maturation of regenerating myelinated axons in mice lacking neurofilaments. Exper Neurol 1997; 148:299-316.
30. Elder GA, Friedrich Jr VL, Bosco P et al. Absence of the mid-sized neurofilament subunit decreases axonal calibers, levels of light neurofilament (NF-L), and neurofilament content. J Cell Biol 1998; 141(3):727-39.
31. Jacomy H, Zhu Q, Couillard-Despres S et al. Disruption of type IV intermediate filament network in mice lacking the neurofilament medium and heavy subunits. J Neurochem 1999; 73(3):972-84.
32. Xu Z, Marszalek JR, Lee MK et al. Subunit composition of neurofilaments specifies axonal diameter. J Cell Biol 1996; 133(5):1061-9.
33. Rao MV, Campbell J, Yuan A et al. The neurofilament middle molecular mass subunit carboxyl-terminal tail domains is essential for the radial growth and cytoskeletal architecture of axons but not for regulating neurofilament transport rate. J Cell Biol 2003; 163(5):1021-31.
34. Levavasseur F, Zhu Q, Julien JP. No requirement of alpha-internexin for nervous system development and for radial growth of axons. Brain Res Mol Brain Res 1999; 69(1):104-12.
35. Lariviere RC, Nguyen MD, Ribeiro-da-Silva A et al. Reduced number of unmyelinated sensory axons in peripherin null mice. J Neurochem 2002; 81(3):525-32.
36. McGraw TS, Mickle JP, Shaw G et al. Axonally transported peripheral signals regulate alpha-internexin expression in regenerating motoneurons. J Neurosci 2002; 22(12):4955-63.
37. Elder GA, Friedrich Jr VL, Margita A et al. Age-related atrophy of motor axons in mice deficient in the mid-sized neurofilament subunit. J Cell Biol 1999; 146(1):181-92.
38. Hirano A, Nakano I, Kurland LT et al. Fine structural study of neurofibrillary changes in a family with amyotrophic lateral sclerosis. J Neuropathol Exp Neurol 1984; 43(5):471-80.
39. Galloway PG, Mulvihill P, Perry G. Filaments of Lewy bodies contain insoluble cytoskeletal elements. Am J Pathol 1992; 140(4):809-22.
40. Munoz DG, Greene C, Perl DP et al. Accumulation of phosphorylated neurofilaments in anterior horn motoneurons of amyotrophic lateral sclerosis patients. J Neuropathol Exp Neurol 1988; 47(1):9-18.
41. Corbo M, Hays AP. Peripherin and neurofilament protein coexist in spinal spheroids of motor neuron disease. J Neuropathol Exp Neurol 1992; 51(5):531-7.
42. Cote F, Collard JF, Julien JP. Progressive neuronopathy in transgenic mice expressing the human neurofilament heavy gene: A mouse model of amyotrophic lateral sclerosis. Cell 1993; 73(1):35-46.
43. Xu Z, Cork LC, Griffin JW et al. Increased expression of neurofilament subunit NF-L produces morphological alterations that resemble the pathology of human motor neuron disease. Cell 1993; 73(1):23-33.
44. Eyer J, Peterson A. Neurofilament-deficient axons and perikaryal aggregates in viable transgenic mice expressing a neurofilament-beta-galactosidase fusion protein. Neuron 1994; 12(2):389-405.
45. Lee MK, Marszalek JR, Cleveland DW. A mutant neurofilament subunit causes massive, selective motor neuron death: Implications for the pathogenesis of human motor neuron disease. Neuron 1994; 13(4):975-88.
46. Wong PC, Marszalek J, Crawford TO et al. Increasing neurofilament subunit NF-M expression reduces axonal NF-H, inhibits radial growth, and results in neurofilamentous accumulation in motor neurons. J Cell Biol 1995; 130(6):1413-22.
47. Marszalek JR, Williamson TL, Lee MK et al. Neurofilament subunit NF-H modulates axonal diameter by selectively slowing neurofilament transport. J Cell Biol 1996; 135(3):711-24.
48. Gama Sosa MA, Friedrich Jr VL, DeGasperi R et al. Human midsized neurofilament subunit induces motor neuron disease in transgenic mice. Exp Neurol 2003; 184(1):408-19.
49. Lee VM, Otvos Jr L, Carden MJ et al. Identification of the major multiphosphorylation site in mammalian neurofilaments. Proc Natl Acad Sci USA 1988; 85(6):1998-2002.
50. Shneidman PS, Carden MJ, Lees JF et al. The structure of the largest murine neurofilament protein (NF-H) as revealed by cDNA and genomic sequences. Brain Res 1988; 464(3):217-31.
51. Beaulieu JM, Nguyen MD, Julien JP. Late onset death of motor neurons in mice overexpressing wild-type peripherin. J Cell Biol 1999; 147(3):531-44.
52. Ching GY, Chien CL, Flores R et al. Overexpression of alpha-internexin causes abnormal neurofilamentous accumulations and motor coordination deficits in transgenic mice. J Neurosci 1999; 19(8):2974-86.
53. Hand CK, Rouleau GA. Familial amyotrophic lateral sclerosis. Muscle Nerve 2002; 25(2):135-59.
54. Julien JP, Beaulieu JM. Cytoskeletal abnormalities in amyotrophic lateral sclerosis: Beneficial or detrimental effects? J Neurol Sci 2000; 180(1-2):7-14.

55. Figlewicz DA, Krizus A, Martinoli MG et al. Variants of the heavy neurofilament subunit are associated with the development of amyotrophic lateral sclerosis. Hum Mol Genet 1994; 3(10):1757-61.
56. Tomkins J, Usher P, Slade JY et al. Novel insertion in the KSP region of the neurofilament heavy gene in amyotrophic lateral sclerosis (ALS). Neuroreport 1998; 9(17):3967-70.
57. Al-Chalabi A, Andersen PM, Nilsson P et al. Deletions of the heavy neurofilament subunit tail in amyotrophic lateral sclerosis. Hum Mol Genet 1999; 8(2):157-64.
58. Berger P, Young P, Suter U. Molecular cell biology of Charcot-Marie-Tooth disease. Neurogenetics 2002; 4(1):1-15.
59. Mersiyanova IV, Perepelov AV, Polyakov AV et al. A new variant of Charcot-marie-tooth disease type 2 is probably the result of a mutation in the neurofilament-light gene. Am J Hum Genet 2000; 67(1):37-46.
60. De Jonghe P, Mersivanova I, Nelis E et al. Further evidence that neurofilament light chain gene mutations can cause Charcot-Marie-Tooth disease type 2E. Ann Neurol 2001; 49(2):245-9.
61. Georgiou DM, Zidar J, Korosec M et al. A novel NF-L mutation Pro22Ser is associated with CMT2 in a large Slovenian family. Neurogenetics 2002; 4(2):93-6.
62. Yoshihara T, Yamamoto M, Hattori N et al. Identification of novel sequence variants in the neurofilament-light gene in a Japanese population: Analysis of Charcot-Marie-Tooth disease patients and normal individuals. J Peripher Nerv Syst 2002; 7(4):221-4.
63. Jordanova A, De Jonghe P, Boerkoel CF et al. Mutations in the neurofilament light chain gene (NEFL) cause early onset severe Charcot-Marie-Tooth disease. Brain 2003; 126(Pt 3):590-7.
64. Dauer W, Przedborski S. Parkinson's disease: Mechanisms and models. Neuron 2003; 39(6):889-909.
65. Lavedan C, Buchholtz S, Nussbaum RL et al. A mutation in the human neurofilament M gene in Parkinson's disease that suggests a role for the cytoskeleton in neuronal degeneration. Neurosci Lett 2002; 322(1):57-61.
66. Kruger R, Fischer C, Schulte T et al. Mutation analysis of the neurofilament M gene in Parkinson's disease. Neurosci Lett 2003; 351(2):125-9.
67. Bergeron C, Beric-Maskarel K, Muntasser S et al. Neurofilament light and polyadenylated mRNA levels are decreased in amyotrophic lateral sclerosis motor neurons. J Neuropathol Exp Neurol 1994; 53(3):221-30.
68. Wong NK, He BP, Strong MJ. Characterization of neuronal intermediate filament protein expression in cervical spinal motor neurons in sporadic amyotrophic lateral sclerosis (ALS). J Neuropathol Exp Neurol 2000; 59(11):972-82.
69. Menzies FM, Grierson AJ, Cookson MR et al. Selective loss of neurofilament expression in Cu/Zn superoxide dismutase (SOD1) linked amyotrophic lateral sclerosis. J Neurochem 2002; 82(5):1118-28.
70. Robertson J, Doroudchi MM, Nguyen MD et al. A neurotoxic peripherin splice variant in a mouse model of ALS. J Cell Biol 2003; 160(6):939-49.
71. Canete-Soler R, Schwartz ML, Hua Y et al. Stability determinants are localized to the 3'-untranslated region and 3'-coding region of the neurofilament light subunit mRNA using a tetracycline-inducible promoter. J Biol Chem 1998; 273(20):12650-4.
72. Canete-Soler R, Schwartz ML, Hua Y et al. Characterization of ribonucleoprotein complexes and their binding sites on the neurofilament light subunit mRNA. J Biol Chem 1998; 273(20):12655-61.
73. Canete-Soler R, Schlaepfer WW. Similar poly(C)-sensitive RNA-binding complexes regulate the stability of the heavy and light neurofilament mRNAs. Brain Res 2000; 867(1-2):265-79.
74. Canete-Soler R, Silberg DG, Gershon MD. et al. Mutation in neurofilament transgene implicates RNA processing in the pathogenesis of neurodegenerative disease. J Neurosci 1999; 19(4):1273-83.
75. Nie Z, Wu J, Zhai J et al. Untranslated element in neurofilament mRNA has neuropathic effect on motor neurons of transgenic mice. J Neurosci 2002; 22(17):7662-70.
76. Lin H, Zhai J, Nie Z et al. Neurofilament RNA causes neurodegeneration with accumulation of ubiquitinated aggregates in cultured motor neurons. J Neuropathol Exp Neurol 2003; 62(9):936-50.
77. Canete-Soler R, Wu J, Zhai J et al. p190RhoGEF Binds to a destabilizing element in the 3' untranslated region of light neurofilament subunit mRNA and alters the stability of the transcript. J Biol Chem 2001; 276(34):32046-50.
78. Ge W, Wu J, Zhai J et al. Binding of p190RhoGEF to a destabilizing element on the light neurofilament mRNA is competed by BC1 RNA. J Biol Chem 2002; 277(45):42701-5.
79. Hadano S, Hand CK, Osuga H et al. A gene encoding a putative GTPase regulator is mutated in familial amyotrophic lateral sclerosis 2. Nat Genet 2001; 29(2):166-73.
80. Yang Y, Hentati A, Deng HX et al. The gene encoding alsin, a protein with three guanine-nucleotide exchange factor domains, is mutated in a form of recessive amyotrophic lateral sclerosis. Nat Genet 2001; 29(2):160-5.

81. Yamanaka K, Vande Velde C, Eymard-Pierre E et al. Unstable mutants in the peripheral endosomal membrane component ALS2 cause early-onset motor neuron disease. Proc Natl Acad Sci USA 2003; 100(26):16041-6.
82. Ge WW, Leystra-Lantz C, Wen W et al. Selective loss of trans-acting instability determinants of neurofilament mRNA in amyotrophic lateral sclerosis spinal cord. J Biol Chem 2003; 278(29):26558-63.
83. Nissim-Rafinia M, Kerem B. Splicing regulation as a potential genetic modifier. Trends Genet 2002; 18(3):123-7.
84. Stoilov P, Meshorer E, Gencheva M et al. Defects in pre-mRNA processing as causes of and predisposition to diseases. DNA Cell Biol 2002; 21(11):803-18.
85. Faustino NA, Cooper TA. Pre-mRNA splicing and human disease. Genes Dev 2003; 17(4):419-37.
86. Sternberger LA, Sternberger NH. Monoclonal antibodies distinguish phosphorylated and nonphosphorylated forms of neurofilaments in situ. Proc Natl Acad Sci USA 1983; 80(19):6126-30.
87. Lee VM, Carden MJ, Schlaepfer WW et al. Monoclonal antibodies distinguish several differentially phosphorylated states of the two largest rat neurofilament subunits (NF-H and NF-M) and demonstrate their existence in the normal nervous system of adult rats. J Neurosci 1987; 7(11):3474-88.
88. Brown A. Contiguous phosphorylated and non-phosphorylated domains along axonal neurofilaments. J Cell Sci 1998; 111(Pt 4):455-67.
89. Gotow T, Tanaka J, Takeda M. The organization of neurofilaments accumulated in perikaryon following aluminum administration: Relationship between structure and phosphorylation of neurofilaments. Neuroscience 1995; 64(2):553-69.
90. Sobue G, Hashizume Y, Yasuda T et al. Phosphorylated high molecular weight neurofilament protein in lower motor neurons in amyotrophic lateral sclerosis and other neurodegenerative diseases involving ventral horn cells. Acta Neuropathol (Berl) 1990; 79(4):402-8.
91. Schmidt ML, Carden MJ, Lee VM et.al. Phosphate dependent and independent neurofilament epitopes in the axonal swellings of patients with motor neuron disease and controls. Lab Invest 1987; 56(3):282-94.
92. Collard JF, Cote F, Julien JP. Defective axonal transport in a transgenic mouse model of amyotrophic lateral sclerosis. Nature 1995; 375(6526):61-4.
93. Williamson TL, Cleveland DW. Slowing of axonal transport is a very early event in the toxicity of ALS-linked SOD1 mutants to motor neurons. Nat Neurosci 1999; 2(1):50-6.
94. Zhang B, Tu P, Abtahian F et al. Neurofilaments and orthograde transport are reduced in ventral root axons of transgenic mice that express human SOD1 with a G93A mutation. J Cell Biol 1997; 139(5):1307-15.
95. Hirokawa N. Cross-linker system between neurofilaments, microtubules, and membranous organelles in frog axons revealed by the quick-freeze, deep- etching method. J Cell Biol 1982; 94(1):129-42.
96. Morris RL, Hollenbeck PJ. Axonal transport of mitochondria along microtubules and F-actin in living vertebrate neurons. J Cell Biol 1995; 131(5):1315-26.
97. Hollenbeck PJ. The pattern and mechanism of mitochondrial transport in axons. Front Biosci 1996; 1:d91-d102.
98. Wagner OI, LIFshitz J, Janmey PA et.al. Mechanisms of mitochondria-neurofilament interactions. J Neurosci 2003; 23(27):9046-58.
99. Leterrier JF, Rusakov DA, Nelson BD et.al. Interactions between brain mitochondria and cytoskeleton: Evidence for specialized outer membrane domains involved in the association of cytoskeleton-associated proteins to mitochondria in situ and in vitro. Microsc Res Tech 1994; 27(3):233-61.
100. Bajaj NP, Miller CC. Phosphorylation of neurofilament heavy-chain side-arm fragments by cyclin-dependent kinase-5 and glycogen synthase kinase-3alpha in transfected cells. J Neurochem 1997; 69(2):737-43.
101. Sun D, Leung CL, Liem RKH. Phosphorylation of the high molecular weight neurofilament protein (NF- H) by Cdk5 and p35. J Biol Chem 1996; 271(24):14245-51.
102. Veeranna, Amin ND, Ahn NG et al. Mitogen-activated protein kinases (Erk1,2) phosphorylate Lys-Ser-Pro (KSP) repeats in neurofilament proteins NF-H and NF-M. J Neurosci 1998; 18(11):4008-21.
103. Brownlees J, Yates A, Bajaj NP et al. Phosphorylation of neurofilament heavy chain side-arms by stress activated protein kinase-1b/Jun N-terminal kinase-3. J Cell Sci 2000; 113(Pt 3):401-7.
104. Link WT, Grant P, Hidaka H et al. Casein kinase I and II from squid brain exhibit selective neurofilament phosphorylation. Mol Cell Neurosci 1992; 3:548-58.
105. Hollander BA, Bennett GS, Shaw G. Localization of sites in the tail domain of the middle molecular mass neurofilament subunit phosphorylated by a neurofilament-associated kinase and by casein kinase I. J Neurochem 1996; 66(1):412-20.
106. Nakamura Y, Hashimoto R, Kashiwagi Y et al. Casein kinase II is responsible for phosphorylation of NF-L at Ser-473. FEBS Lett 1999; 455(1-2):83-6.

107. Hisanaga S, Matsuoka Y, Nishizawa K et al. Phosphorylation of native and reassembled neurofilaments composed of NF-L, NF-M, and NF-H by the catalytic subunit of cAMP-dependent protein kinase. Mol Biol Cell 1994; 5(2):161-72.

108. Cleverley KE, Betts JC, Blackstock WP et al. Identification of novel in vitro PKA phosphorylation sites on the low and middle molecular mass neurofilament subunits by mass spectrometry. Biochemistry 1998; 37(11):3917-30.

109. Yoshimura Y, Aoi C, Yamauchi T. Investigation of protein substrates of Ca(2+)/calmodulin-dependent protein kinase II translocated to the postsynaptic density. Brain Res Mol Brain Res 2000; 81(1-2):118-28.

110. Tanaka J, Ogawara M, Ando S et al. Phosphorylation of a 62 kd porcine alpha-internexin, a newly identified intermediate filament protein. Biochem Biophys Res Commun 1993; 196(1):115-23.

111. Aletta JM, Angeletti R, Liem RK et al. Relationship between the nerve growth factor-regulated clone 73 gene product and the 58-kilodalton neuronal intermediate filament protein (peripherin). J Neurochem 1988; 51(4):1317-20.

112. Huc C, Escurat M, Djabali K et al. Phosphorylation of peripherin, an intermediate filament protein, in mouse neuroblastoma NIE 115 cell line and in sympathetic neurons. Biochem Biophys Res Commun 1989; 160(2):772-9.

113. Angelastro JM, Ho CL, Frappier T et al. Peripherin is tyrosine-phosphorylated at its carboxyl-terminal tyrosine. J Neurochem 1998; 70(2):540-9.

114. Saito T, Shima H, Osawa Y et al. Neurofilament-associated protein phosphatase 2A: Its possible role in preserving neurofilaments in filamentous states. Biochemistry 1995; 34(22):7376-84.

115. Veeranna, Shetty KT, Link WT et al. Neuronal cyclin-dependent kinase-5 phosphorylation sites in neurofilament protein (NF-H) are dephosphorylated by protein phosphatase 2A. J Neurochem 1995; 64(6):2681-90.

116. Terry-Lorenzo RT, Inoue M, Connor JH et al. Neurofilament-L is a protein phosphatase-1-binding protein associated with neuronal plasma membrane and post-synaptic density. J Biol Chem 2000; 275(4):2439-46.

117. Nakagawa T, Chen J, Zhang Z et al. Two distinct functions of the carboxyl-terminal tail domain of NF-M upon neurofilament assembly: Cross-bridge formation and longitudinal elongation of filaments. J Cell Biol 1995; 129(2):411-29.

118. Gou JP, Gotow T, Janmey PA et al. Regulation of neurofilament interactions in vitro by natural and synthetic polypeptides sharing Lys-Ser-Pro sequences with the heavy neurofilament subunit NF-H: Neurofilament crossbridging by antiparallel sidearm overlapping. Med Biol Eng Comput 1998; 36(3):371-87.

119. Leterrier JF, Kas J, Hartwig J et al. Mechanical effects of neurofilament cross-bridges. Modulation by phosphorylation, lipids, and interactions with F-actin. J Biol Chem 1996; 271(26):15687-94.

120. Gonda Y, Nishizawa K, Ando S et al. Involvement of protein kinase C in the regulation of assembly-disassembly of neurofilaments in vitro. Biochem Biophys Res Commun 1990; 167(3):1316-25.

121. Hisanaga S, Yasugawa S, Yamakawa T et al. Dephosphorylation of microtubule-binding sites at the neurofilament-H tail domain by alkaline, acid, and protein phosphatases. J Biochem (Tokyo) 1993; 113(6):705-9.

122. Miyasaka H, Okabe S, Ishiguro K et al. Interaction of the tail domain of high molecular weight subunits of neurofilaments with the COOH-terminal region of tubulin and its regulation by tau protein kinase II. J Biol Chem 1993; 268(30):22695-702.

123. Jung C, Shea TB. Regulation of neurofilament axonal transport by phosphorylation in optic axons in situ. Cell Motil Cytoskeleton 1999; 42(3):230-40.

124. Pant HC. Dephosphorylation of neurofilament proteins enhances their susceptibility to degradation by calpain. Biochem J 1988; 256(2):665-8.

125. Kampfl A, Zhao X, Whitson JS et al. Calpain inhibitors protect against depolarization-induced neurofilament protein loss of septo-hippocampal neurons in culture. Eur J Neurosci 1996; 8(2):344-52.

126. Nixon RA, Paskevich PA, Sihag RK et al. Phosphorylation on carboxyl terminus domains of neurofilament proteins in retinal ganglion cell neurons in vivo: Influences on regional neurofilament accumulation, interneurofilament spacing, and axon caliber. J Cell Biol 1994; 126(4):1031-46.

127. de Waegh SM, Lee VM, Brady ST. Local modulation of neurofilament phosphorylation, axonal caliber, and slow axonal transport by myelinating Schwann cells. Cell 1992; 68(3):451-63.

128. Sanchez I, Hassinger L, Sihag RK et al. Local control of neurofilament accumulation during radial growth of myelinating axons in vivo. Selective role of site-specific phosphorylation. J Cell Biol 2000; 151(5):1013-24.

129. Hsieh ST, Kidd GJ, Crawford TO et al. Regional modulation of neurofilament organization by myelination in normal axons. J Neurosci 1994; 14(11 Pt 1):6392-401.

130. Tortarolo M, Veglianese P, Calvaresi N et al. Persistent activation of p38 mitogen-activated protein kinase in a mouse model of familial amyotrophic lateral sclerosis correlates with disease progression. Mol Cell Neurosci 2003; 23(2):180-92.

131. Atzori C, Ghetti B, Piva R et al. Activation of the JNK/p38 pathway occurs in diseases characterized by tau protein pathology and is related to tau phosphorylation but not to apoptosis. J Neuropathol Exp Neurol 2001; 60(12):1190-7.

132. Nguyen MD, Lariviere RC, Julien JP. Deregulation of Cdk5 in a mouse model of ALS: Toxicity alleviated by perikaryal neurofilament inclusions. Neuron 2001; 30(1):135-47.

133. Bajaj NP, Al-Sarraj ST, Anderson V et al. Cyclin-dependent kinase-5 is associated with lipofuscin in motor neurones in amyotrophic lateral sclerosis. Neurosci Lett 1998; 245(1):45-8.

134. Pei JJ, Grundke-Iqbal I, Iqbal K et al. Accumulation of cyclin-dependent kinase 5 (cdk5) in neurons with early stages of Alzheimer's disease neurofibrillary degeneration. Brain Res 1998; 797(2):267-77.

135. Ahlijanian MK, Barrezueta NX, Williams RD et al. Hyperphosphorylated tau and neurofilament and cytoskeletal disruptions in mice overexpressing human p25, an activator of cdk5. Proc Natl Acad Sci USA 2000; 97(6):2910-5.

136. Patrick GN, Zukerberg L, Nikolic M et al. Conversion of p35 to p25 deregulates Cdk5 activity and promotes neurodegeneration. Nature 1999; 402(6762):615-22.

137. Ferri A, Gabbianelli R, Casciati A et al. Calcineurin activity is regulated both by redox compounds and by mutant familial amyotrophic lateral sclerosis-superoxide dismutase. J Neurochem 2000; 75(2):606-13.

138. Previtali SC, Zerega B, Sherman DL et al. Myotubularin-related 2 protein phosphatase and neurofilament light chain protein, both mutated in CMT neuropathies, interact in peripheral nerve. Hum Mol Genet 2003; 12(14):1713-23.

139. Nixon RA. Dynamic behavior and organization of cytoskeletal proteins in neurons: Reconciling old and new findings. Bioessays 1998; 20(10):798-807.

140. Shea TB. Microtubule motors, phosphorylation and axonal transport of neurofilaments. J Neurocytol 2000; 29(11-12):873-87.

141. Almenar-Queralt A, Goldstein LS. Linkers, packages and pathways: New concepts in axonal transport. Curr Opin Neurobiol 2001; 11(5):550-7.

142. Brown A. Slow axonal transport: Stop and go traffic in the axon. Nat Rev Mol Cell Biol 2000; 1(2):153-6.

143. Yuan A, Rao MV, Kumar A et al. Neurofilament transport in vivo minimally requires hetero-oligomer formation. J Neurosci 2003; 23(28):9452-8.

144. Gurney ME, Pu H, Chiu AY et al. Motor neuron degeneration in mice that express a human Cu,Zn superoxide dismutase mutation. Science 1994; 264(5166):1772-5.

145. Tu PH, Raju P, Robinson KA et al. Transgenic mice carrying a human mutant superoxide dismutase transgene develop neuronal cytoskeletal pathology resembling human amyotrophic lateral sclerosis lesions. Proc Natl Acad Sci USA 1996; 93(7):3155-60.

146. Borchelt DR, Wong PC, Becher MW et al. Axonal transport of mutant superoxide dismutase 1 and focal axonal abnormalities in the proximal axons of transgenic mice. Neurobiol Dis 1998; 5(1):27-35.

147. Warita H, Itoyama Y, Abe K. Selective impairment of fast anterograde axonal transport in the peripheral nerves of asymptomatic transgenic mice with a G93A mutant SOD1 gene. Brain Res 1999; 819(1-2):120-31.

148. Xia CH, Roberts EA, Her LS et al. Abnormal neurofilament transport caused by targeted disruption of neuronal kinesin heavy chain KIF5A. J Cell Biol 2003; 161(1):55-66.

149. Brownlees J, Ackerley S, Grierson AJ et al. Charcot-marie-tooth disease neurofilament mutations disrupt neurofilament assembly and axonal transport. Hum Mol Genet 2002; 11(23):2837-44.

150. Puls I, Jonnakuty C, LaMonte BH et al. Mutant dynactin in motor neuron disease. Nat Genet 2003; 33(4):455-6.

151. LaMonte BH, Wallace KE, Holloway BA et al. Disruption of dynein/dynactin inhibits axonal transport in motor neurons causing late-onset progressive degeneration. Neuron 2002; 34(5):715-27.

152. Hafezparast M, Klocke R, Ruhrberg C et al. Mutations in dynein link motor neuron degeneration to defects in retrograde transport. Science 2003; 300(5620):808-12.

153. Couillard-Despres S, Zhu Q, Wong PC et al. Protective effect of neurofilament heavy gene overexpression in motor neuron disease induced by mutant superoxide dismutase. Proc Natl Acad Sci USA 1998; 95(16):9626-30.

154. Williamson TL, Bruijn LI, Zhu Q et al. Absence of neurofilaments reduces the selective vulnerability of motor neurons and slows disease caused by a familial amyotrophic lateral sclerosis-linked superoxide dismutase 1 mutant. Proc Natl Acad Sci USA 1998; 95(16):9631-6.

155. Nguyen MD, Boudreau M, Kriz J et al. Cell cycle regulators in the neuronal death pathway of amyotrophic lateral sclerosis caused by mutant superoxide dismutase 1. J Neurosci 2003; 23(6):2131-40.

CHAPTER 4

Neurofilaments:
Phosphorylation and Signal Transduction

Sashi Kesavapany, Richard H. Quarles and Harish C. Pant*

Abstract

Neurofilaments belong to the Class IV family of Intermediate filaments and are neuron-specific. They are classed into three distinct groups according to their molecular masses; NF-H (heavy chain), NF-M (middle-chain) and NF-L (light chain). Together with microtubules and their associated proteins, neurofilaments make up the dynamic axonal cytoskeleton. Neurofilaments comprise a central alpha helical coil-coiled domain flanked by an amino terminal head domain and in the case of NF-H and NF-M, a hypervariable carboxy-terminal tail domain. Neurofilaments participate in dynamic properties of the axonal cytoskeleton such as axon outgrowth, axonal transport and the control of axonal caliber. They contain multiple phosphorylation sites in their amino-head and carboxy-tail domains that are phosphorylated topographically by a number of kinases and the phosphorylation of neurofilaments is related to their functions. Expression and phosphorylation of neurofilaments is developmentally regulated and most of the phosphorylation occurs in the tail domains in the mature nervous system. Kinases that have been found to phosphorylate neurofilaments include PKA, CKI, CKII CaMK and the proline directed kinases. The proline directed kinases such as Cdk5, members of the MAP kinase family (p38, SAPK and ERK1/2) and GSK3 phosphorylate the KSP repeats found in NF-M and NF-H tail domains. Aberrant hyperphosphorylation of neurofilaments in dendrites and cell bodies are seen in neurodegenerative diseases such as Amyotrophic Lateral Sclerosis, Alzheimer's disease, Parkinson's disease, Pick's disease and Dementia with Lewy bodies. Thus, defects in compartmentalization of cytoskeletal protein phosphorylation may contribute to pathology seen in these diseases. Neurofilament phosphorylation is affected by signal transduction pathways. Calcium influx into neurons causes the phosphorylation of NF-M through the activation of the ERK1/2. Integrin mediated signaling also causes the phosphorylation of NF-H through the activation of Cdk5 activity. Recent studies have also shown that kinase cascades can be affected by myelin associated glycoprotein (MAG), a major glial protein found in periaxonal membranes of glial cells. MAG appears to be involved in bi-directional signaling affecting axonal properties such as axonal caliber, phosphorylation of neurofilaments and mediating the activity of ERK1/2 and Cdk5. Thus, signal transduction pathways are involved in the phosphorylation of the KSP repeats found in NF-M and NF-H.

*Corresponding Author: Harish Pant—Room 4D-28, Building 36, MSC 4130, NINDS, NIH, 36 Convent Drive, Bethesda, Maryland 20892, U.S.A. Email: panth@ninds.nih.gov

Intermediate Filaments, edited by Jesus Paramio. ©2006 Landes Bioscience and Springer Science+Business Media.

Neurofilaments Are Members of the Intermediate Filament Family

Neurofilaments belong to the Class IV family of Intermediate Filaments (IF) and are the neuron specific intermediate filaments.[1-3] Vertebrate neurofilaments are approximately 10 nm in diameter and are intermediate in size between microfilaments and microtubules. They consist of three groups divided according to their molecular mass; neurofilament heavy-chain (NF-H: 190 kDa), neurofilament middle-chain (NF-M: 115 kDa) and neurofilament light-chain (NF-L: 68 kDa). In common with other members of the intermediate filament family, NF-H, NF-M and NF-L each comprise a central alpha helical coil-coiled rod domain flanked by a variant amino-terminal globular head domain and a hyper-variable carboxy-terminal tail domain which differ in length among the sub-units[1,4,5] (see Fig. 1).

Together with microtubules, microtubule-associated proteins (MAPs) 1A, 1B and tau, actin and other associated proteins, neurofilaments (NFs) make up the dynamic axonal cytoskeleton. Of the cytoskeletal components, only the neurofilaments are unique to and diagnostic of neurons. In some large axons (eg. squid giant axons), NF proteins make up to 13% of the total protein.[6,7] This suggests that NFs play an important structural role in neurons and together with MAPs and other associated proteins, they sustain axonal and dendritic branching patterns and also promote axonal growth or thickening. It is now clear that NFs also participate in dynamic properties of the axonal cytoskeleton during neuronal differentiation, axon outgrowth and guidance and its state of phosphorylation seems to be important with regards to its function.[6,7]

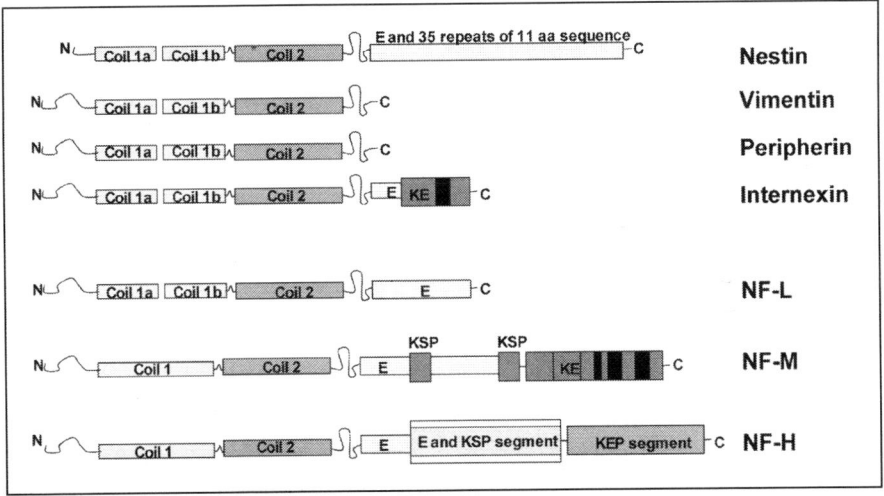

Figure 1. Domains of neuronal intermediate filaments (adapted from Nixon and Shea, 1992). Coil 1 and coil 2 regions are conserved alpha-helical domains containing heptad repeats of hydrophobic amino acids. These regions are separated by nonhelical "links" which are represented by the separation of the boxes. NF-M and NF-H contain a unique, single elongated coil 1 region, which may restrict their ability to form filaments in the absence of a 'backbone' subunit such as NF-L. The initial twist of the carboxy-terminus, which constitutes the tail region of vimentin and peripherin, is followed by the specialized domains in the other neuronal intermediate filaments species as follows. The carboxy-terminus of nestin contain glutamic acid-rich segments (E) and approximately 35 repeats of an 11 amino acid sequence. Nestin also contains a relatively short amino-terminus. The neurofilaments, along with internexin, contain one or more glutamic acid-rich segments while internexin possesses a region enriched in lysine and glutamic acid (KE). NF-M and NF-H both contain repeating regions of lysine-serine-proline (KSP) segments which represent the regions that are extensively phosphorylated in these two species. The extreme termini of NF-M and NF-H possess unique regions rich in lysine and glutamic acid (KE segment) and in lysine, glutamic acid and proline (KEP) residues. The black boxes within the KE region of NF-M represent unique amino acid repeats which may be involved in binding functions.

The central rod domain is involved in coiled-coil formation of the filamentous structure while the globular amino head domain is involved in neurofilament assembly. The head domain of neurofilaments contain multiple second-messenger related phosphorylation sites which may regulate head domain function in assembly of neurofilaments.[8-11] The carboxy-terminal tail domains are distinct in NF- H and NF- M. The carboxy-terminal extensions form side-arms that extend from the filament and appear to form connections between adjacent neurofilaments and between neurofilaments and other axoplasmic structures. The most striking feature of NF-M and NF-H is the carboxy-terminal (C-terminal) side arm domains which contain multiple lysine/serine/proline (KSP) repeat motifs. These are known to be phosphorylated in axons in vivo. NF-H is one of the most extensively phosphorylated neuronal proteins containing over 50 potential acceptor sites (8-100 depending on the species).[12] Mass spectroscopic analyses of NF-H in humans, rats, canine and squid have shown that most of the KSP sites are phosphorylated in vivo.[13,14]

Developmentally, the neuronal cytoskeleton must accommodate the morphological and behavioral transitions from migrating and target-seeking axons to the stability of a functioning, mature neuron. NFs are phenotypic markers of differentiating neurons and prior to their synthesis, neuronal precursor molecules express, either independently or together with NFs, other intermediate filament (IF) proteins such as nestin, vimentin, α-internexin and peripherin.[2] Although the patterns of expression of proteins within the developing nervous system exhibits temporal and spatial differences, the timing of different intermediate filament expression show a common trend. One of the first proteins to be expressed during neurogenesis is nestin, which is a ClassVI IF protein and is expressed in the neuroepithelia of the developing neural tube. As neuroblasts replicate within the neuroepithelium, they also express vimentin, another IF protein which may be involved in cell proliferation at this stage.[15,16] It may also promote neurite outgrowth in the neural progenitor cells since anti-sense vimentin constructs inhibited this process in these cells.[17,18] In other neurons with small axons, peripherin or α-internexin are expressed during this same early period of neurogenesis. Peripherin is homologous to Class III IF proteins and its expression correlates with the outgrowth of specific CNS and PNS neuronal populations. Its expression is transient in most CNS neurons and it maintains plasticity during the axon outgrowth phase and gradually disappears as NF subunits are synthesized to form the stable cytoskeleton. However, peripherin appears to be expressed more in some PNS neurons.[18] It is expressed into adulthood in dorsal root ganglia neurons (DRGs) and in regenerating adult DRG neurons, initial neurite regrowths are enriched in peripherin. The peripherin expression also disappears as neurofilaments are expressed. Although peripherin can form filaments with NF-L in transfected cells, it cannot assemble with NF-M and NF-H subunits.[19] Early expression of α-internexin in CNS neurons correlates with an increase in axonal outgrowth persisting in adult neurons. This may implicate a role in assembly with neurofilaments, however, this seems to be unimportant in neurogenesis, since an α-internexin null-mouse exhibits a completely normal phenotype, without any changes in axon trajectories, connectivity and caliber.[20] Neurofilaments, for the most part, exhibit a sequential order of expression in vertebrate development with NF-L and NF-M appearing initially as neurites differentiate. Later, as the nervous system matures, NF-H expression becomes apparent in the brain and spinal cord, as stabilization of the neuronal cytoskeleton occurs. Thus, at embryonic day 12 (E12) in the rat, the first post-mitotic regions in the hindbrain begin to express NF-L and then shortly thereafter NF-M. These NFs are probably heterodimers of NF-M/L, are unphosphorylated and their function includes the maintenance of the cytoskeletal plasticity for neuronal migration and axon outgrowth. These neurons, containing the unphosphorylated NFs are considered immature and at this early stage, peripherin and α-internexin are also present and may colocalize with NFs in some neurons. This two stage expression pattern for IFs and NFs also occur in other cell types such as cultured E18 hippocampal neurons, E15 DRG neurons, and adult regenerating neurons of the rat and *Xenopus*.[18,21-23] The pattern of mRNA expression also mirrors the protein expression profile with NF-L and NF-M mRNA being expressed days before the expression of NF-H mRNA.[24] Thus, it appears that the expression of NF-L and NF-M genes are coordinately regulated while the NF-H gene is expressed independently. A similar pattern of NF expression is seen in the developing squid nervous system, however, one alternatively spliced gene gives rise to all three NF proteins in this system. In all these studies though, it is important to bear in mind that the detection of the neurofilament proteins are heavily dependent on the epitope specificity of the antibodies used, particularly the phospho-specific ones and in some cases such antibodies were not available to facilitate a thorough development study. The synthesis of IF proteins (including NFs), during neurogenesis is initially detected in neuronal cell bodies in the earliest stages of differentiation. Subsequently, NFs begin to be detected in neurites possibly due to axonal transport. This topographic pattern was observed using PC12 cells in the presence and in the absence of nerve growth factor (NGF). In the presence of NGF, neurite outgrowth is observed in these cells, however, in the absence of NGF, the cells contain low levels of all three NF proteins and any neurofilaments are detected only in the perikarya of the cell.[25,26] Although filaments are detected in the cell, these were considered incomplete or diassociated filaments however, after NGF induction, the levels of NF protein expression markedly increased, particularly within neurites.

Neurofilaments and Axonal Caliber

One of the functions of neurofilaments is to control the axonal caliber since the speed of conductivity of a nerve impulse along the axon is directly proportional to the caliber of the axon.[27-29] Neurofilaments are particularly abundant in large myelinated axons such as those of motor neurons where speed is crucial in their proper function. Neurofilament involvement in caliber determination was demonstrated through analyses of a mutant 'quiver' quail in which the NF-L gene was disrupted and thus neurofilament formation was affected.[30] The diameter of axons was significantly reduced in the mutant 'quiver' quail as was the conduction velocity of impulses. Similar studies have now been confirmed in similar models in transgenic mice.[31,32]

Phosphorylation of Neurofilaments

After their transport into the axons, neurofilaments are extensively phosphorylated.[2,29,33-35] In the developmental paradigm, phosphorylated NF-M can be detected as early as E13 in spinal cord neurons whereby its expression is most intense in the axons and peripheral nerves with little or no expression in the perikarya and dendritic branches. NF-M and NF-L expression reach peak, adult values at E15, where NF-H is detected in brain and spinal cord at low levels and is a mixture of phospho and nonphospho forms. Most of the neurofilament phosphorylation sites are located in the KSP motifs of the tail domains of NF-M and NF-H, although other Ser/Thr motifs are phosphorylated.[36] The spatio-temporal phosphorylation pattern of neurofilaments is distinct in different neuronal sub-populations and cerebellar granule cell parallel fibers, which lack NF-H, show no phosphorylation.[33] On the other hand, in the neurons of the optic nerve, NF phosphorylation occurs in a gradient function along the axon, starting proximally as neurofilaments enter the axon hillock and continuing down the axon.[2] Nonphosphorylated NF-M/H is found in cell bodies of DRG, ventral motor and sympathetic neurons while phosphorylated NF-M/H is found down the axon suggesting the phosphorylation state of NF-M/H is related to the extent of axonal transport of neurofilaments.[37] However, this matter is still debated since phospho-NF-H is intensified in brain and spinal cord neurons but its phosphorylation pattern occurs distal-proximal (ie. opposite to the other neurons). The suggestion here is that perhaps contact with terminals initiate NF phosphorylation.[38] The phosphorylation of NFs continue postnatally in different brain regions at different rates and in some regions such as the olfactory bulb and hypothalamus, phospho-NF-H is undetectable at postnatal day 1 (P1).[39] The initial appearance and progressive phosphorylation of NF-M and NF-H in the axons are region-specific and correlated to synaptogenesis and myelination, as the mature neuronal cytoskeleton begins to be established.[40] Myelination also signals to promote NF phosphorylation and radial growth or thickening of the axons (described later). Interestingly, NF phosphorylation is not seen in embryonic or adult neuronal dendrites even though all three subunits are present.[33,41] This may be due to a lack of specific signals to activate the kinases responsible for NF subunit phosphorylation. The axonal phosphorylation of NFs is a slow process in vivo and may rely on the subunits being transported into the axons however, the transport of NFs is slowed down by their phosphorylation. This may account for the phospho-gradients observed along the axon however, other factors such as the regulation of the kinase activities and conformation of the proteins are also important determinants in NF phosphorylation.

NF-L is essential for the formation of neurofilaments from its subunits, however, the precise mechanism by which neurofilaments form is controversial. Neither NF-M nor NF-H can self assemble into filaments in vitro and the central α-helical rod domain of NF-L controls NF formation by acting as the backbone.[42] Interestingly, NF-L can self assemble into 10 nm homopolymeric filaments in vitro.[43] In vitro, NF-L assembly probably begins with the dimers assembling into tetrameric protofilaments which then form octameric protofibrils and ultimately into core 10nm wide filaments.[44] Both ends of NF-L were found to act antagonistically and were required for NF-L assembly in vitro, by experiments where tail-less, head-less and rod-less constructs of NF-L were recombined in various combinations. The head domain promotes lateral association of protofibrils into protofilaments and the tail domain promotes the termination of filaments upon reaching 10 nm.[44] In vivo however, NFs are obligate heteropolymers although NF-L is required for NF formation, it is not essential for viablility as observed by extensive transgenic modulation and gene knockout studies. The mutation of the NF-L gene in the Japanese quail results in the production of a fertile, viable adult with a 'quiver' phenotype. The 'quiver' quail exhibits NF-L deficient axons without neurofilaments decreased axonal caliber and earlier death than the wild-type neurons. A NF-L null mouse does not have any neurofilaments, exhibits a reduced axonal caliber but this mutant also contains dramatic reductions in NF-M and NF-H levels.[30] In transient transfection studies using cells that lack intermediate filaments, NF-H and NF-M were unable to form homopolymeric filaments whereas NF-L, cotransfected with NF-M or NF-H, assembled into filamentous structures. Experiments that removed the tail domains of NF-L and NF-M still allowed coassembly into filaments thus indicating that the tail domains are indispensable for filament formation even in vitro.[45] Instead, another region of neurofilaments were found to be important in for filament formation. The assembly into a heteropolymer is dependent on the head domains of NF-L and NF-M and more specifically on the phosphorylation states of these domains. The phosphorylation of the head domain of NF-L inhibits filament assembly or it may promote disassembly. Protein kinase A (PKA) mediated phosphorylation of Serine[55] in the head domain of NF-L or Serine[44] in the head domain of NF-M is sufficient to block assembly.[46-48] Additionally, transgenic mice expressing a mutant NF-L, Serine[55] to Aspartate[55] mutation (mimicking permanent phosphorylation), exhibited pathological accumulations of neurofilament aggregates in the cell bodies of neurons within a month after birth.[49] Thus, the phosphorylation of the head domains may occur immediately after NF-L synthesis in the cell bodies so that premature assembly of neurofilaments in the cell bodies is averted, before transport into the axons.

Neurofilaments and Their Associated Kinases

Intermediate filaments such as vimentin, desmin and GFAP are phosphorylated by PKA and PKC. Much like vimentin, the head domain of NF-L can be phosphorylated by PKA. This phosphorylation prevents NF-L assembly and induces the disassembly of NF-L filaments in vitro. The major PKA target site on NF-L is localized in the head domain on the Ser[55] residue, a site that may be essential for early stages of regulation of filament assembly.[49] Assembly or disassembly of NF-L filaments in vitro is also regulated by phosphorylation of Ser[57] in

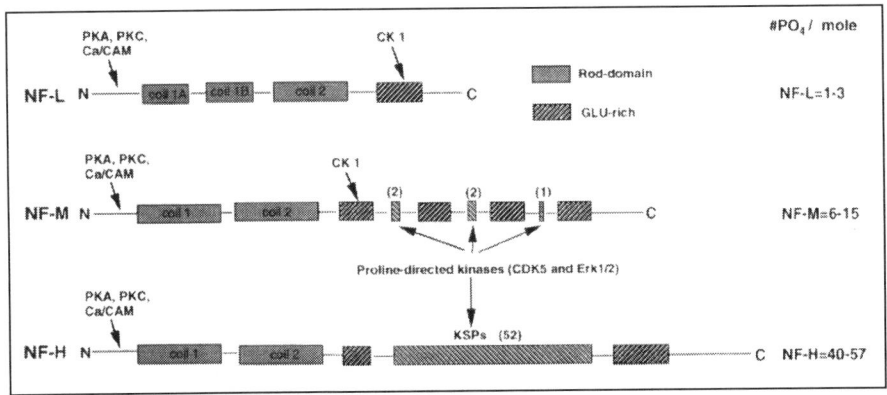

Figure 2. Neurofilament subunit domains and sites that are phosphorylated by known kinases. Schematic representation of NF-L, NF-M and NF-H proteins and sites that are phosphorylated by different kinases. All three subunits are phosphorylated at their amino termini by protein kinase A (PKA), protein kinase C (PKC) and calcium-calmodulin kinase (CA/CAM). Both NF-L and NF-M are phosphorylated in the Glutamic acid rich region (shaded box) by Casein kinase I (CKI). NF-M and NF-H are phosphorylated in their KSP repeat regions by ERK1/2 and Cdk5. The number of phosphates incorporated per mole of neurofilament protein are indicated on the right. Nf-L and NF-M have 1-3 and 6-15 phosphates per mole respectively and NF-H has between 40-57 phosphates incorporated all seemingly attributed to ERK and Cdk5.

the head domain of NF-L by the Rho associated kinases.[50] Second messenger independent kinases such as CKI and CKII are found in NF preparations from mammalian nervous tissue. CKI, a constitutively active kinase, is the principal kinase associated with NFs in preparations made from squid axoplasm, bovine and chicken spinal cord[51-53] and in vitro, CKI phosphoylates Ser/Thr residues in the C-terminal tail domain shared by all three subunits.[46,51-55] CKI phosphoylates at least 5 Ser residues C-terminal to a conserved VEEIIEET sequence found in all three NF subunits from human to lamprey.[56] CKII, a microtubule associated kinase that phosphoylates β-tubulin,[57,58] is also associated with NF preparations. It is less active than CKI with NF substrates, particularly NF-H, but it phosphoylates NF-L at Ser[473] in vitro. Additionally, an antibody that specifically detects phospho-NF-L exhibited robust expression in perikarya of rat cortex suggesting this event takes place in vivo too.[59] CAMKII, another kinase enriched in squid NF preparations phosphoylates NF proteins among other substrates in the complex of PSD proteins[60] (see Fig. 2).

Neurofilament Tail Domain Phosphorylation

The C-terminal domains of all neurofilament proteins contain a glutamic acid rich region with serine residues. These serine residues can be phosphorylated by casein kinase I and II on the basis of their consensus sequences.[46,51,52,61]

Most phosphorylation sites in NF-M and NF-H are found in their KSP repeat regions, which can be specifically phosphorylated by the proline directed kinases. The tail domains of the NF-M and NF-H contain consensus sequences of Cdk5 and the mitogen activated protein kinase (MAPK) family. The kinases associated with NFs isolated from neuronal tissue have been extensively analysed. These preparations contain endogenous kinase activities that phosphoylate a number of substrates including NFs. The kinases isolated in these preparations phosphoylate casein and lysine rich histone thus suggesting the presence of multiple kinases in the NF preparation. It is now evident that two families of the proline directed kinases, the cyclin dependent kinases such as Cdk5 and the MAP kinases such as ERK1/2, SAP, p38 kinases and GSK3 are the principal kinases that phosphoylate the KSP repeats in the NF-M and NF-H tail domains[62-73] (see Fig. 2). NF-H is the most extensively phosphorylated molecule in the nervous system and most of the phosphorylation takes place in the C-terminal KSP repeats.[62-73] In vivo phosphorylation of NF-H by Cdk5 has been shown to occur in transfection studies. Cotransfection studies of NF-H with p35/Cdk5 or GSK in COS cells, have shown that NF-H is phosphorylated at different sites by the two kinases. Only p35/Cdk5 altered human NF-H electrophoretic mobility to a similar level to brain NF-H suggesting that KSPXK repeats were phosphorylated by Cdk5.[73] Rodent NF-H contains 41 KSPXXXK repeats between residues 508 and 763, which were not phosphorylated by p35/Cdk5. The specificity of Cdk5 for the KSPXK was confirmed in another study where the KSPXXXK motif, which predominates in the tail domain of rat NF-H is preferentially phosphorylated by ERK1/2 while Cdk5, which requires an adjacent basic or neutral residue, has a greater preference for KSPXK.[73,74] This explains why human NF-H, with its 34 KSPXK repeats is a preferred substrate for Cdk5.

Axonal Transport of Neurofilaments

Synthesis of all the cytoskeleton molecules, their kinases and regulators occurs in the perikarya of the neuron and delivered by a mechanism called axonal transport (see for review refs. 75-78). Axonal transport is divided into two major categories based on the speed of transport. These are 'fast axonal transport' which includes membraneous components and 'slow axonal transport' that includes cytoskeletal components such as neurofilaments, microtubules and their associated proteins. Initially it was thought that the different rates of transport were due to the transported proteins associating with different motors responsible for the two types of transport. However, it is also possible that the different speeds are due to the proteins attaching, for varying lengths of time, onto a unitary motor. Thus, elements undergoing fast transport could be attached to the motor all the time therefore being transported rapidly and elements undergoing slow transport, could be attached to the motor for a fraction of the time. Motors that could translocate material in both anterograde and retrograde directions along microtubules have been identified but the same success has not been obtained when identifying motors responsible for slow axonal transport. The known retrograde fast transport motor, dynein, was shown to mediate the movement of microtubules and similarly, the fast anterograde motor kinesin transported NFs in cellular system and in situ.[79-81] Studies have also implicated dynein in movement of NFs along MTs.[82] Real time experiments have shown that NF transport in cultured neurons that what appears to be what was first thought to be slow transport of NFs is actually a series of short 'sprints' by the NFs at speeds consistent with fast axonal transport, mixed with periods of nonmotility.[83,84] In addition to NFs being transported as filaments, some NF subunits are transported as punctuate structures consistent at slow transport rates and these structures are then converted into filamentous structures along the length of axonal neurites. Additionally, a smaller percentage of punctuate structures moved in a retrograde fashion, some changed between anterograde and retrograde and some showed no movement.[80,85] NF transport is dependent on the presence of MTs and MT motors. The microinjection of anti-kinesin antibodies disrupted NF transport and furthermore, NFs are recovered in a MT motor preparation and are coimmunoprecipitated by anti-kinesin antibodies.[79] NF retrograde transport has also been documented in peripheral axon ligations in mice where both bulk and newly transported NFs accumulated on both proximal and distal ends of the ligated nerve. The proximal accumulations could be explained by anterograde transport while the distal end accumulations could only be explained by retrograde transport. Mechanical constriction of axons of cultured neurons yielded similar results.[86] Real time or 'live' experiments, using GFP tagged NF-H constructs, showed that a proportion of NF-H moved in a predominantly retrograde fashion with occasional anterograde movement while a fraction moved in an opposite fashion.[80,83-85] Recent studies suggest that dynein may mediate the retrograde movement of NFs. NFs were observed to undergo expected bi-directional movement and copurified with dynein and kinesin-related motors. NF motility was predominantly towards an anterograde direction when antibodies against the cargo-loading domain of dynein were introduced, suggesting that dynein may be responsible for the retrograde movement.[82]

Neurofilament phosphorylation is implicated in the overall slowing of their transport rate as migration along the axon continues. The rate of NF transport is inversely correlated with their phosphorylation state. Hypophosphorylated NF-H and NF-M are transported approximately twice as fast as their phosphorylated forms. Furthermore, NF-H, containing a phospho-epitope normally seen later in development, is transported at half the rate of total NF-H.[87,88] The reversible association of NFs with their transport complexes could be regulated by phosphorylation. Increased C-terminal phosphorylation may promote disassociation of NFs from their anterograde transport motor. Studies manipulating the kinases and phosphatases also support a role for phosphorylation in the regulation of NF transport.[79,89] The injection of the phosphatase inhibitor, okadaic acid, slowed axonal transport of NFs and also induced the appearance of a phospho-NF epitope within the proximal segments of optic axons that is normally restricted to distal areas. The slowing of NF transport was not due to an overall inhibition of axonal transport since another component, fodrin, was not affected.[90] The relationship between the phosphorylation event, the kinases resposnsible and association of NFs with the transport machinery is highly complex and not completely understood. If a phosphatase inhibitor inhibits transport, then it follows that a kinase inhibitor may then increase transport, but this is not the case. PD98059, a specific inhibitor of the MAP kinase pathway inhibits the anterograde transport of GFP-tagged NF subunits in neuroblastoma cells however, mitochondrial transport was not affected suggesting that not all anterograde transport was impaired. Additionally, olomoucine, an inhibitor of Cdk5 inhibited both mitochondrial as well as NF transport.[91] Further experiments showed the complexity of the transport mechanism where, as mentioned above, cells treated with PD98059, suppressed NF transport but, addition of MAP kinase to a microtubule motor preparation disrupted the interaction between NFs and kinesin while at the same time increasing C-terminal NF phosphorylation.[81] Recent studies have shown that NF phosphorylation is responsible for disassociating NFs from kinesin and hypophosphorylated NFs were recovered from a standard MT motor preparation rich in kinesin.[81] The failure to recover kinesin from routine NF preparations could be because the majority of NFs are highly phosphorylated. When MAP kinase is added to the motor preparation, there was an increase in C-terminal phosphorylation of some of the NF subunits and these were unable to coprecipitate with kinesin.[81] This selective phosphorylation and subsequent dissociation from the anterograde motor may be a potential mechanism by which NF phosphorylation is implicated in the slowing of NF axonal transport.[81,92] Ongoing work to determine which of the approximately 50 phospho-sites of NFs regulate the association of NFs to the motors and whether different kinases, phosphorylating different sites have divergent effects remain critical areas of investigation.

Phosphorylation of NFs and Disease

The KSP sites in NF-M and NF-H are extensively phosphorylated in axons but not in dendrites or cell bodies except in cases of neurodegenerative diseases such as Amyotrophic Lateral Sclerosis (ALS) where abnormally phosphorylated NFs accumulate in the spinal motor neuron perikarya.[93] A number of neurodegenerative diseases such as Alzheimer's, Pick's disease with Lewy bodies, Dementia with Lewy bodies and Parkinson's disease are characterized by a common neuronal pathology where there are accumulations of insoluble filamentous aggregates with neuronal perikarya (see for review ref. 94). Thus, defects in the compartmentalization of cytoskeletal protein phosphorylation may lead to neuronal death seen in these diseases. The principal proteins found in the aggregates are tau and synuclein, but small amounts of phosphorylated neurofilaments and their associated kinases have also been detected through immunodetection of the proteins. ALS however, is a different story as the major pathological hallmark is neurofilaments containing filamentous aggregates primarily localized to the perikarya and the proximal axons of spinal motor neurons.[93] Additionally, the neurofilament aggregates are extensively phosphorylated, particularly in the tail domains of NF-M and NF-H. Neurons that contain this pathology are destined to die accounting for the motor neuron loss, nerve degeneration and the muscle dystrophy that is characteristic of ALS, resulting in death. Even though the pathological hallmark has been discovered, the precise mechanism of neuronal loss in ALS is confusing and not fully understood yet. Studies attempting to shed some light on the mechanism have involved altering the neurofilament protein subunit stoichiometry with neurons using transgenic mice and a number of different models have been produced. However these models did not explain what was going wrong in the neurons, as diverse and contradictory neuronal and behavioural phenotypes were obtained. A 2-4 fold overexpression of either human or mouse NF-H produces a massive accumulation of NFs within spinal motor neuron perikarya but only overexpression of human NF-H produces overt motor neuron dysfunction characteristic of the disease.[95-98] Like-wise, transgenic mice over-expressing NF-L also display a spinal neuron pathology resembling that of ALS with phosphorylated NFs in the perikarya.[99,100] Over expression of mutant or wild-type murine NF-M or human NF-M leads to some perikaryal accumulations within motor neurons however in the majority of cases, massive accumulations, motor dysfunction nor neuronal loss were not observed.[31,101-103] Studies performed with NF-M knock-out mice exhibited decreases in NF-L protein levels while NF-H protein levels and NF transport velocity increase suggesting that NF-M plays a dominant role in the regulation of NF-L protein stoichiometry and the phosphorylation state of NF-H.[104,105] Transgenic NF-M mice have increased levels of NF-L and both total and phospho-NF-H levels decrease as if to compensate for the overexpression of NF-M. These studies suggest that as long as the NF-L to NF-H ratio is maintained, filaments are assembled and transported as normal. As confusing as the studies are, the stoichiometric imbalance among the NF subunits are primarily responsible for the spinal motor neuron phenotype seen in ALS since introduction of NF-L in to NF-H transgenic mice restore a normal phenotype.[106] Neurons with higher levels of axonal NFs seem to be most at risk from the imbalances in stoichiometry of the NF subunits since transgenic mice overexpressing peripherin did not display any massive accumulations in the spinal motor neurons but some had significant accumulations in the proximal regions, accompanied by motor dysfunction from 6-28 months.[19,98] To confuse the situation even further, the nature of the aggregates in the motor neurons are not identical, with some being spherical and some filamentous and the nature of associated proteins unknown. Even the state of NF phosphorylation is not the predictor for neuronal cell death as, a recent comparison of the specific pattern of NF-H phosphorylation between ALS and normal spinal cords were similar.[92] So the question remains about what factors are responsible for neuronal cell death, axon degeneration and the motor dysfunction seen in ALS. It appears that changes in the NF subunit ratio in the cell body may affect NF assembly, NF transport and/or NF phosphorylation or perhaps all three in some combination or altogether.[92,107] A common feature of all the affected neurons in the mouse models is the serious disruption of NF transport, perhaps due to the premature assembly of abnormal polymers within the cell bodies and this inhibits proper association with the microtubule transport machinery.[92,94,97,107] The phosphorylation of NF-M and NF-H tail domains may also contribute to the disruption of NF transport due to the mechanical interference cause by the side-arm formation. Consistent with this observation, transgenic mice bearing superoxide-1 (SOD1) mutations display motor neuron accumulations of phosphorylated neurofilaments.[108] Autosomal dominant mutations in the SOD-1 gene are the primary cause of ALS in a small number of human families. Transgenic mice with these mutations exhibit an early onset of reduced tubulin transport and subsequent appearance of motor neuron pathology and cell death characteristic of ALS.

Neurofilaments and Signal Transduction Pathways

Only a small subset of Ser/Thr residues in rat NF-H are phosphorylated by Cdk5. The MAP kinases, particularly ERK1/2, play a greater role in the phosphorylation of these residues.[12] In vivo, the MAP kinase cascade is implicated in neurite outgrowth in PC12 cells where NGF stimulation induces a sustained increase in ERK1/2 activation and neurite outgrowth is

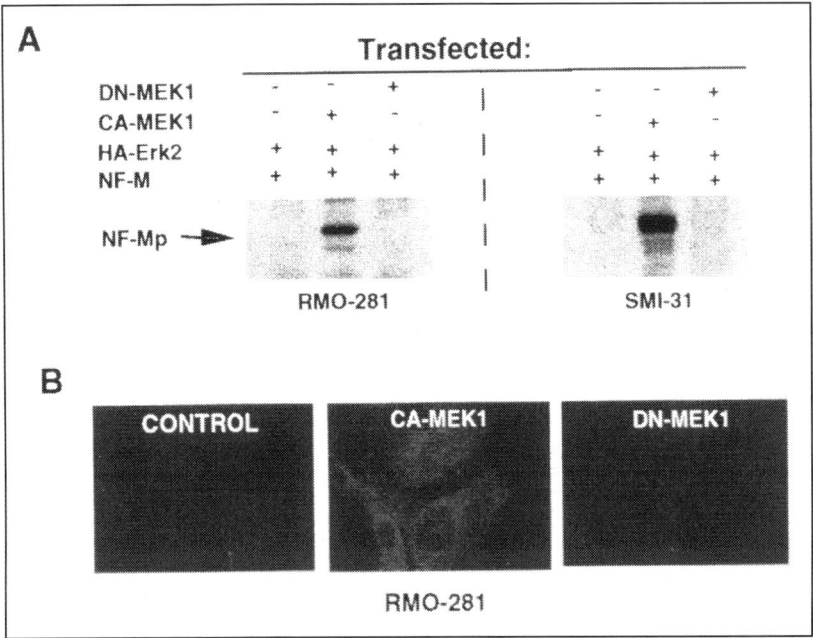

Figure 3. The transient transfection with the constitutively active MEK1 construct causes the phosphorylation of NF-M tail domain. A) NIH3T3 cells were transiently transfected with NF-M and hemagluttinin tagged (HA)-ERK2 (lane 1), NF-M, HA-ERK2 and CA-MEK1 (lane 2) and NF-M, HA-ERK2 and dominant negative (DN)-MEK1 (lane 3). Phospho NF-M was detected from cell lysates through Western blot analysis, using the monoclonal phospho-specific antibodies RMO-281 (left panel) and SMI31 (right panel). Equal amounts of protein were loaded and resolved in each lane. B) The same set of transfections were subjected to immunocytochemical analyses. Transfected, serum-starved cells were fixed and incubated with RMO-281, followed by FITC-conjugated secondary antibody. It was clear that transfections with CA-MEK1 caused the phosphorylation of NF-M and DN-MEK1 did not induce NF-M phosphorylation.

detected within a few days.[109] Accompanying this neurite outgrowth is the activation of NF-L and NF-M mRNA transcription while NF-H transcription is unchanged.[25] Inhibition of the MAP kinase pathway results in the inhibition of neurite outgrowth and in hippocampal cells, this is accompanied by a decrease NF-M, NF-H and MAP-1 phosphorylation, suggesting a role for these proteins in axon elongation.[73] The MAP kinase signal cascade is also implicated in calcium-induced membrane depolarization of PC12 cells. The induction of calcium influx through L-type channels resulted in the activation of ERK1/2 phosphorylation and subsequent neurite outgrowth and phosphorylation of the NF-M tail domain.[68] Inhibition of the calcium influx or the MAP kinase pathway with the use of a specific MEK1 (upstream activator of ERK1/2) inhibitor, prevented neurite outgrowth and NF-M phosphorylation. Additionally, constitutively active MEK1, cotransfected with a NF-M tail domain construct in NIH 3T3 cells, resulted in ERK1/2 activation and the NF-M tail domain phosphorylation (Fig. 3). Further experiments involving NIH3T3 cells treated with EGF, a growth factor that specifically activates the MAP kinase pathway and transfected with NF-M showed an activation of ERK1/2 and induced NF-H phosphorylation.[69] This suggested that endogenous MEK1 was activated and the effects were blocked by the MEK1 inhibitor suggesting that the MAP kinase cascade is involved in regulating NF tail domain phosphorylation. It has been shown that ERK1/2 phosphoylates the KSP repeats in NF-M and NF-H and this event, is linked to neurite outgrowth and branching in primary rat hippocampal neurons. Treatment of these cells with the MEK1 inhibitor PD98059 caused a decrease in outgrowth and inhibited the

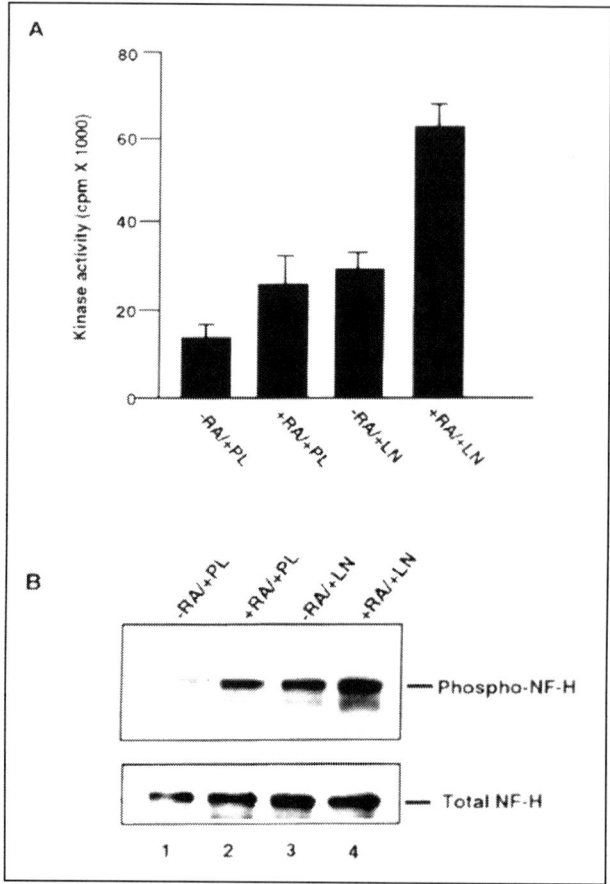

Figure 4. Retinoic acid and laminin induced Cdk5 activity and NF-H KSP tail domain phosphorylation in SH-SY-5Y human neuroblastoma cells. A) Cells were treated with retinoic acid (RA-10 μM) or with vehicle for 7 days and then culured on poly-L-lysine or laminin for 24 hours. Cell extracts were immunoprecipated with anti-Cdk5 antibody (c8) and immunoprecipitates were assayed for their ability to phoshorylate histone H1. IN the presence of retinoic acid and laminin kinase activity was observed to be highest. B) Equal amounts of protein from cell lysates were immunoblotted using total (SMI33) and phospho-dependent NF-H (SMI31) antibodies. Again, total NF-H levels were barely changed but phosphorylation of NF-H increased dramatically in the presence of laminin.

phosphorylation of NF-H and NF-M and MAP. Additionally, studies have shown that the phosphorylation event is regulated by signal transduction pathways. Recent studies have shown that integrin α1β1 activated Cdk5 and this causes the phosphorylation of NF-H[110] (Fig. 4). Cellular adhesion to the extracellular matrix is mediated by a diverse class of α1/β1 heterodimeric receptors known as integrins, which transduce signals at the cell surface to activate multiple intracellular signal pathways within the cells. The signaling pathways that link integrins to neuronal process outgrowth is not fully understood. It was discovered that the α1β1 integrin induces the activation of Cdk5, causing phosphorylation of human C-terminal NF-H on KSP repeats as well as mediating neurite outgrowth. These results indicated that Cdk5 may play an important role in promoting neurite outgrowth and human NF-H tail domain phosphorylation through the α1β1 integrin signaling pathway (Fig. 5).

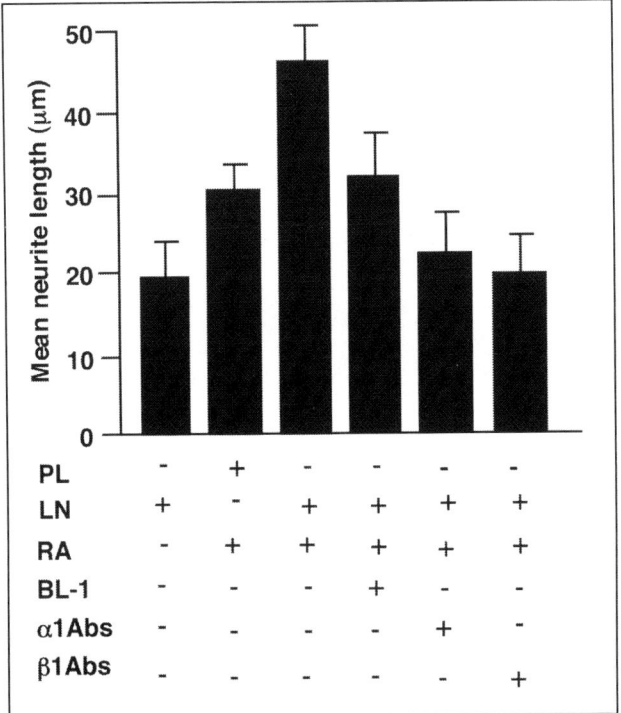

Figure 5. Laminin induced Cdk5 activity correlated with neurite outgrowth in RA-treated SH-SY-5Y cells. SH-SY-5Y cells were evaluated for neurite outgrowth in the presence and absence of α1β1 antibodies or the Cdk5 inhibitor BL-1. Cells were treated with and without RA for 7 days, detached and replated on poly-L-lysine or laminin coated coverslips in serum-free media with and without α1 antibodies, β1 antibodies and BL-1. Processes were measure in five randomly selected fields after fixation and staining with crystal-violet. Laminin and RA together caused a larger increase in neurite outgrowth compared with laminin or RA alone. α1 and β1 antibodies and BL-1 inhibited this effect. The antibodies appeared to be more effective in reducing neurite outgrowth compared to BL-1.

The phosphorylation of neurofilaments occurs in axons in close proximity to myelin sheaths.[111] Myelination may be the signal needed to induce phosphorylation of NFs in axons but it is also possible that a signal from Schwann cells or oligodendrocytes may also activate Cdk5 and MAP kinase. Increased phosphorylation of MAPs are also caused by the presence of myelin sheaths on axons and thus, myelination may cause the increased axonal caliber primarily determined by the phosphorylation of NFs. However, the specific molecules in myelinating cells and mechanisms present, that regulate signaling cascades affecting the expression and phosphorylation of cytoskeletal elements, have not been fully identified. The myelin-associated glycoprotein (MAG) belongs to the 'siglec' subgroup of the immunoglobulin superfamily and is expressed exclusively by myelinating oligodendrocytes and Schwann cells.[112,113] Its localization in periaxonal membranes suggests that it is involved in glia/axon interactions and/or signaling and MAG appears to be involved in bi-directional signalling and may be a potential neuronal ligand that modulates glial events. Although studies have shown that MAG is not essential for myelin formation, MAG null mice display subtle abnormalities including the disruption of the periaxonal junction, delayed myelination and formation of dystrophic oligodendrocytes.[114-117] MAG may be particularly important in neuron to glia signaling with MAG acting as a ligand for a neuronal receptor to affect axonal properties. Aging MAG null mice exhibit PNS axonal degeneration, characterized by a reduced

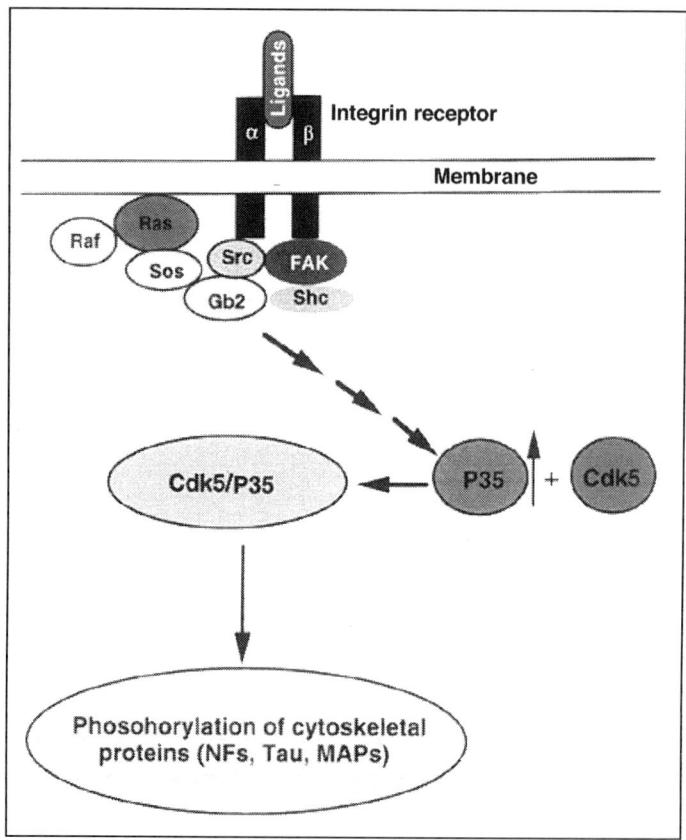

Figure 6. Schematic representaion of the integrin signalling pathway. The membrane bound integrin receptor couple to intracellular signaling molecules, binds to a ligand and is transduced to activate Src-FAK tyrosine kinases. These kinases are linked to the Gb2 and Shc proteins that are also known to associate with the RAs signaling proteins. The message is then shuttle into the cell to increase the transcription of p35 thus activating Cdk5 and resulting in the phosphrylation of cytoskeletal proteins such as neurofilaments MAPs and tau.

axonal caliber with decreased spacing, expression and phosphorylation of NFs.[118,119] Additionally, MAG also affects neurite outgrowth via neuronal signaling cascades.[120-123] The in vitro results suggest that the phosphorylation of NFs is increased when DRG or PC12 neurons are cultured in the presence of MAG, reinforcing the hypothesis that MAG itself is a component of the signaling system that affects the neuronal cytoskeleton (Figs. 6, 7). The increased neurofilament packing density correlates with decreased phosphorylation of KSP repeats in the C-terminal domains of NF-M and NF-H. In addition to confirming the decrease of phospho NF-M and NF-H levels in MAG-null mice as reported previously, the activities of 2 proline directed kinases known to phosphoylate NF-M and NF-H, Cdk5 and MAP kinase, are also reduced in these mice[124] (Fig. 7). Thus, the activities of Cdk5 and ERK1/2 appear to be responsible, at least in part, for the decreased neurofilament phosphorylation (Figs. 8, 9). In addition to changes to the NFs, the microtubular cytoskeleton of PNS axons is also affected when normal myelination is disrupted. Changes in the levels and/or the phosphorylation of MAPs including tau, MAP1A and MAP1B were also detected when DRG neurons were cocultured with MAG expressing cells (Fig. 10). This suggests that a MAG mediated signaling process may also be involved in the regulation of

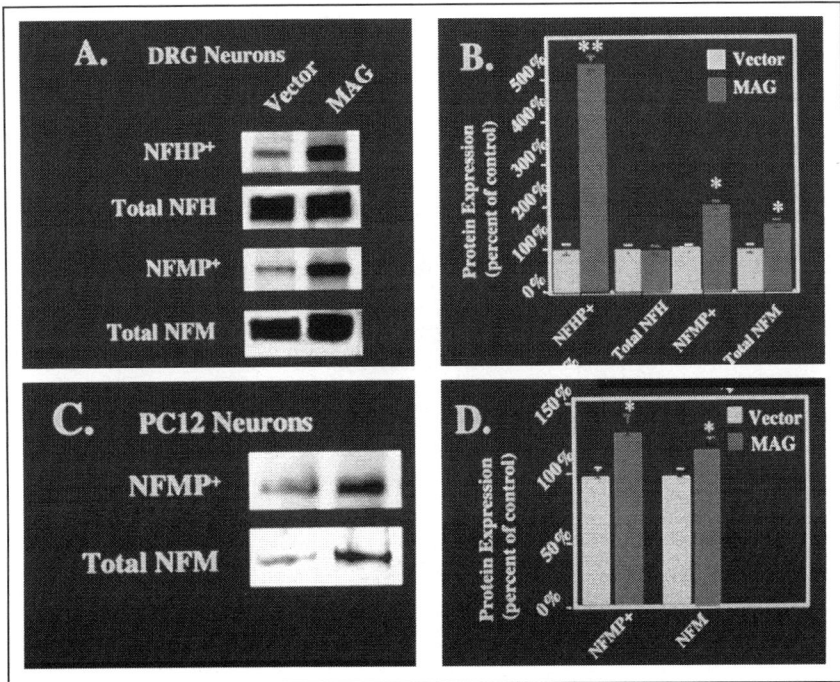

Figure 7. MAG causes the phosphorylation of neurofilaments in DRG neurons and in differentiated PC12 neurons. A) DRG neurons were cocultured with COS cells transfected either with vector only (Vector) or Vector-L-MAG for 4 days. Total protein was analyzed by western blotting and immunodetection of neurofilament subunits were performed using the same antibodies described earlier in Figure 3. Results from three separate experiments are expressed as SEM in the bar graph on the right (B). *p < 0.05, **p < 0.01. C) PC12 neurons were cocultured with transfected either with vector only (Vector) or Vector-L-MAG for 4 days. Western blots showing total and phospho-NFM are shown on the left and results expressed as a bar graph are shown on the right (D). In both cases the presence of MAG stimulated the phosphorylation of NFM without affecting its expression.

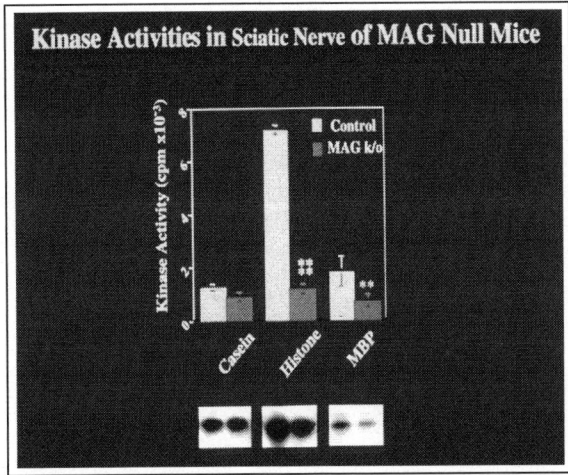

Figure 8. MAG-null Mice have Decreased ERK1/2 and Cdk5 Activity. Sciatic nerves from 10 month old control and MAG-null mice (MAG-/-) were analyzed for ERK1/2 and Cdk5 activity. Phospho-ERK1/2 and Cdk5 were immunoprecipitated and kinase assays using casein, histone and myelin basic protein (MBP) as substrates in in vitro kinase assays. Casein phosphorylation was unchanged between wild-type (Control) and MAG-null mice sciatic nerves. Histone and MBP phosphorylation were significantly reduced. The data are expressed as SEM fro three MAG-null mice and three control (wild-type) mice. *p < 0.05.

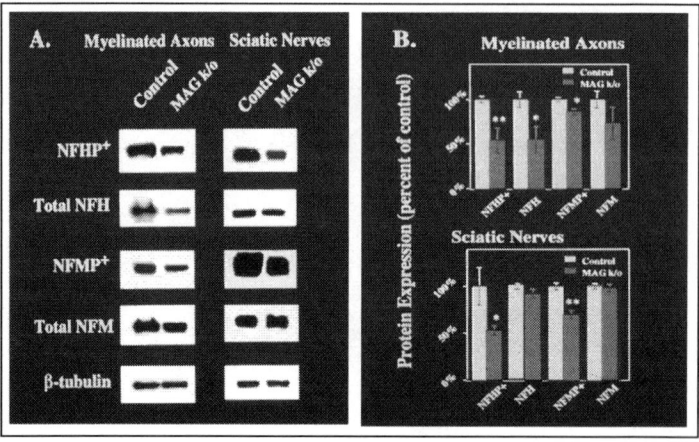

Figure 9. Neurofilament phosphorylation in MAG-null mice. Sciatic nerves from 10 month old control and MAG-null mice (MAG-/-) were analyzed for total and phosphorylated neurofilament proteins by western blotting with the following specific antibodies: SMI31 (NFHP+ and NFMP+), N52 (Total NFH) and RMO270 (Total NFM). NFHP+ and NFMP+ refer to phospho-NFH and NFM respectively. Tubulin is shown as a control that changes were specific to neurofilaments. The bar graph on the right shows results obtained by densitometric quantification of neurofilament proteins in 8 MAG-/- mice expressed as a percentage of mean control levels in 8 control mice (*p < 0.05, **p < 0.01).

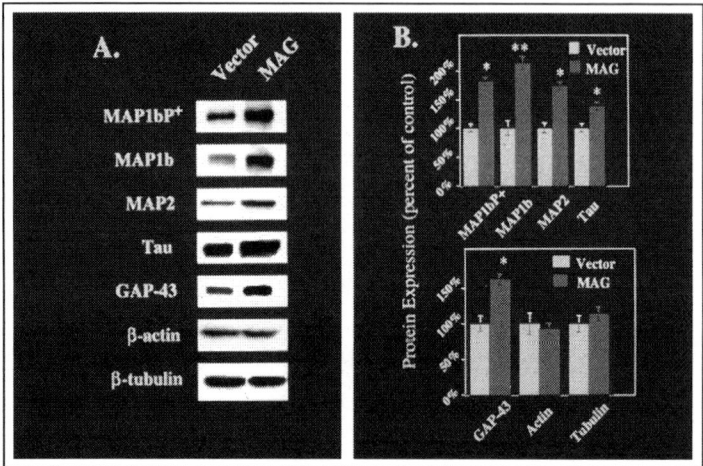

Figure 10. MAG Causes the Increase in MAP2 and Tau Expression and Phosphorylation of MAP1B. DRG neurons were cocultured for 4 days with L-MAG transfected COS cells (MAG) or cells containing plasmid vector only (vector). Following protein extraction, samples were resolved by SDS-PAGE and immunoblotted with the following antibodies: SMI21 to phosphorylated MAP1B (MAP1B+), AA6 (Total MAP1B), MAP2, tau, GAP43, β-actin and β-tubulin. Westerns blots are shown in A. The results of three separate experiments are expressed as mean ± SEM are shown in the bar graphs in B. *p < 0.05. **p < 0.01. MAG expressing COS cells increased both total MAP1B and phosphorylated MAP1B by 2.1 and 1.8 fold respectively, over those DRG neurons containing vector alone. The levels of MAP2 increased 1.7 fold and low molecular weight tau increased 1.4 fold in the presence of MAG. The increased levels of these MAPs in the presence of MAG were associated with a comparable 1.6 fold increase in the expression of growth-associated protein-43, a biochemical marker positively associated with neurite outgrowth. However, there were no significant changes detected in the levels of β-actin and β-tubulin.

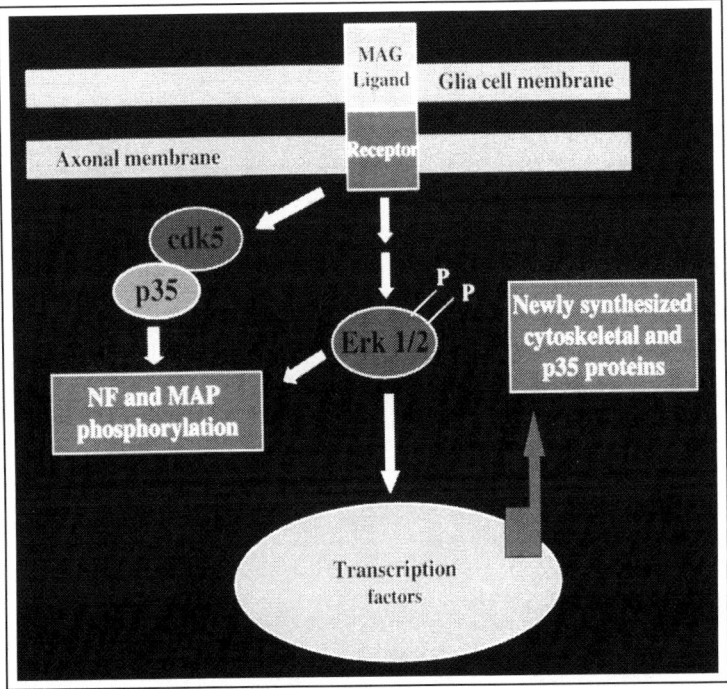

Figure 11. Schematic representation of MAG signaling pathway. The MAG ligand binds to its receptor located in the axonal membrane and initiates a signal transduction pathway involving Cdk5 and ERK1/2 kinases. Its effect on Cdk5/p35 and ERK1/2 kinases results in NF and MAP phosphorylation. ERK1/2 phosphorylation also leads to an increase in transcriptional activation leading to synthesis of cytoskeletal proteins as well as synthesis of the Cdk5 activator, p35.

the microtubular cytoskeleton by myelin. In vitro results strengthen the hypothesis that MAG itself is a component of the Schwann cell periaxonal membrane that signals to the axonal cytoskeleton. The MAG-mediated cellular pathways by which the cytoskeleton may be modified are summarized in Figure 12.

Neurofilaments, MAPS and tau make up the dynamic cytoskeletal architecture of neurons. These proteins are located downstream of initial signals of kinase signal transduction pathways and are phosphorylated by a number of kinases. Of the best characterized kinases, Cdk5 and ERK1/2 have been shown to phosphoylate the cytoskeletal proteins in response to extracellular signals. The various signals and their effects mediate cytoskeletal protein phosphorylation and are summarized in Figures 12 and 13.

Summary

The major questions that remain posed relate to the factors that regulate the compartmentalization of neurofilaments and other cytoskeletal protein phosphorylation in the neurons. Secondly, what are the mechanisms responsible for the regulation and de-regulation seen in neurodegenerative disorders. To address these questions, we have studies the factors regulating the axonal phosphorylation of neurofilament proteins. It is found that the activation of proline-directed kinases, such as ERK1/2 and Cdk5, are primarily responsible for the extensive axonal phosphorylation of multiple KSP repeats within NF-M and NF-H carboxy-tail domains. It has also been demonstrated that these kinases, activated by signal transduction cascades in cell culture systems, phosphorylate NF-M, NF-H and MAPs on their proline-directed

Cell type	Signal	Kinase	Cytoskeletal Proteins
NIH 3T3	EGF	Erk1/2 ⇑	⇑ NF-Mp
PC-12	Ca^{2+}/Membrane Depol.	Erk1/2 ⇑	⇑ NF-Mp
Hippocampal Neurons	MEK inhibition	Erk1/2 ⇓	⇓ NF-M/NF-H, Tau and MAPs Phosphorylation
SH-SY5Y (human Neuroblastoma)	Laminin (Integrin)	Cdk5 ⇑	⇑ NF-Hp
Motor Neurons	Laminin (Integrin)	Erk1/2 ⇑	⇑ NF-Mp

Figure 12. Regulation of cytoskeletal protein phosphorylation by activation and inhibition of kinase cascades. Table showing the effects of signals, the activation of specific kinases and the resultant effect on cytoskeletal protein phosphorylation in various cell types.

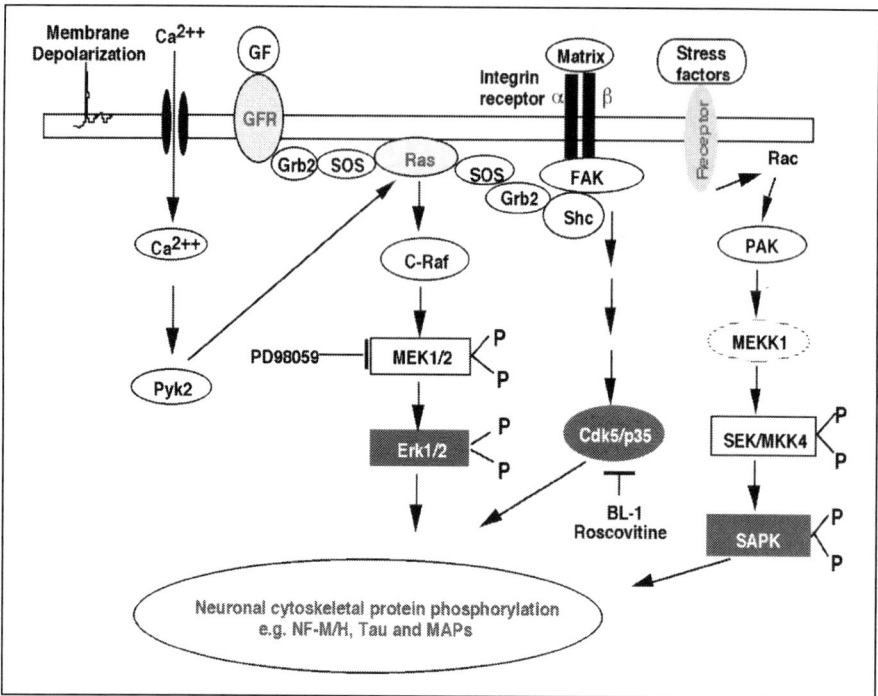

Figure 13. Schematic illustration of neuronal signal transduction pathways leading to cytoskeletal protein phosphorylation. Cellular signals, such as the influx of calcium, leads to the activation of the ERK1/2 pathway through the Ras signaling pathway leading to neuronal cytoskeletal protein phosphorylation. The activation of the Cdk5/p35 kinase complex through as yet unknown steps also leads to the end stage phosphorylation of neurofilaments, tau and MAPs. Stress factors binding to a receptor leads to the activation of the Stress Activated Kinases (SAPK) resulting in cytoskeletal protein phosphorylation. This pathway has not been discussed in this chapter.

Ser or Thr residues. In the axon, one source of exogenous signals are the glia responsible for myelination. Our studies of myelin associated glycoprotein (MAG)-null mice, in which glial cells fail to myelinate axons, have shown that the activities of ERK1/2 and Cdk5, as well as cytoskeletal protein phosphorylation, were down-regulated in the brain tissue of MAG-null mice. To further explore the mechanism of this signaling pathway, we established an in vitro coculture system of PC12 cells or DRG neurons cocultured with MAG-transfected COS cells. Here too we observed activation of ERK1/2 and Cdk5 kinases couple to neurofilament phosphorylation in neuronal cells. In addition, proline directed kinase activation by a wide variety of signal transduction pathways can regulate the cytoskeletal protein phosphorylation in neurons. It is proposed that hyperactivation/deregulation of these kinase cascades may induce aberrant phosphorylation of the cytoskeletal proteins in neurons leading to pathologies seen in neurodegenerative diseases.

References

1. Xu Z, Dong DL, Cleveland DW. Neuronal intermediate filaments: New progress on an old subject. Curr Opin Neurobiol 1994; 4(5):655-661.
2. Nixon RA, Shea TB. Dynamics of neuronal intermediate filaments: A developmental perspective. Cell Motil Cytoskeleton 1992; 22(2):81-91.
3. Pant HC, Floyd CC, Dosemeci A. Neurofilament protein phosphorylation: Topographic regulation and functions. Adv Second Messenger Phosphoprotein Res 1990; 24:393-398.
4. Lee MK, Cleveland DW. Neuronal intermediate filaments. Annu Rev Neurosci 1996; 19:187-217.
5. Wuerker RB, Kirkpatrick JB. Neuronal microtubules, neurofilaments, and microfilaments. Int Rev Cytol 1972; 33:45-75.
6. Morris JR, Lasek RJ. Monomer-polymer equilibria in the axon: Direct measurement of tubulin and actin as polymer and monomer in axoplasm. J Cell Biol 1984; 98(6):2064-2076.
7. Morris JR, Lasek RJ. Stable polymers of the axonal cytoskeleton: The axoplasmic ghost. J Cell Biol 1982; 92(1):192-198.
8. Inagaki M, Gonda Y, Nishizawa K et al. Phosphorylation sites linked to glial filament disassembly in vitro locate in a nonalpha-helical head domain. J Biol Chem 1990; 265(8):4722-4729.
9. Inagaki M, Shibata M. [Regulation of the assembly-disassembly of intermediate filaments]. Tanpakushitsu Kakusan Koso 1989; 34(12 Suppl):1462-1470.
10. Tokutake S. On the assembly mechanism of neurofilaments. Int J Biochem 1990; 22(1):1-6.
11. Kitamura S, Ando S, Shibata M et al. Protein kinase C phosphorylation of desmin at four serine residues within the nonalpha-helical head domain. J Biol Chem 1989; 264(10):5674-5678.
12. Shetty KT, Link WT, Pant HC. cdc2-like kinase from rat spinal cord specifically phosphorylates KSPXK motifs in neurofilament proteins: Isolation and characterization. Proc Natl Acad Sci USA 1993; 90(14):6844-6848.
13. Jaffe H, Veeranna, Shetty KT et al. Characterization of the phosphorylation sites of human high molecular weight neurofilament protein by electrospray ionization tandem mass spectrometry and database searching. Biochemistry 1998; 37(11):3931-3940.
14. Jaffe H, Sharma P, Grant P et al. Characterization of the phosphorylation sites of the squid (Loligo pealei) high-molecular-weight neurofilament protein from giant axon axoplasm. J Neurochem 2001; 76(4):1022-1031.
15. Bennett GS, Tapscott SJ, Kleinbart FA et al. Different proteins associated with 10-nanometer filaments in cultured chick neurons and nonneuronal cells. Science 1981; 212(4494):567-569.
16. Cochard P, Paulin D. Initial expression of neurofilaments and vimentin in the central and peripheral nervous system of the mouse embryo in vivo. J Neurosci 1984; 4(8):2080-2094.
17. Boyne LJ, Fischer I, Shea TB. Role of vimentin in early stages of neuritogenesis in cultured hippocampal neurons. Int J Dev Neurosci 1996; 14(6):739-748.
18. Wong J, Oblinger MM. Differential regulation of peripherin and neurofilament gene expression in regenerating rat DRG neurons. J Neurosci Res 1990; 27(3):332-341.
19. Beaulieu JM, Robertson J, Julien JP. Interactions between peripherin and neurofilaments in cultured cells: Disruption of peripherin assembly by the NF-M and NF-H subunits. Biochem Cell Biol 1999; 77(1):41-45.
20. Levavasseur F, Zhu Q, Julien JP. No requirement of alpha-internexin for nervous system development and for radial growth of axons. Brain Res Mol Brain Res 1999; 69(1):104-112.
21. Athlan ES, Sacher MG, Mushynski WE. Associations between intermediate filament proteins expressed in cultured dorsal root ganglion neurons. J Neurosci Res 1997; 47(3):300-310.

22. Benson DL, Mandell JW, Shaw G et al. Compartmentation of alpha-internexin and neurofilament triplet proteins in cultured hippocampal neurons. J Neurocytol 1996; 25(3):181-196.
23. Zhao Y, Szaro BG. The optic tract and tectal ablation influence the composition of neurofilaments in regenerating optic axons of Xenopus laevis. J Neurosci 1995; 15(6):4629-4640.
24. Breen KC, Anderton BH. Temporal expression of neurofilament polypeptides in differentiating neuroblastoma cells. Neuroreport 1991; 2(1):21-24.
25. Lindenbaum MH, Carbonetto S, Grosveld F et al. Transcriptional and post-transcriptional effects of nerve growth factor on expression of the three neurofilament subunits in PC-12 cells. J Biol Chem 1988; 263(12):5662-5667.
26. Lindenbaum MH, Carbonetto S, Mushynski WE. Nerve growth factor enhances the synthesis, phosphorylation, and metabolic stability of neurofilament proteins in PC12 cells. J Biol Chem 1987; 262(2):605-610.
27. Hoffman PN, Cleveland DW, Griffin JW et al. Neurofilament gene expression: A major determinant of axonal caliber. Proc Natl Acad Sci USA 1987; 84(10):3472-3476.
28. Sakaguchi T, Okada M, Kitamura T et al. Reduced diameter and conduction velocity of myelinated fibers in the sciatic nerve of a neurofilament-deficient mutant quail. Neurosci Lett 1993; 153(1):65-68.
29. Nixon RA, Paskevich PA, Sihag RK et al. Phosphorylation on carboxyl terminus domains of neurofilament proteins in retinal ganglion cell neurons in vivo: Influences on regional neurofilament accumulation, interneurofilament spacing, and axon caliber. J Cell Biol 1994; 126(4):1031-1046.
30. Yamasaki H, Itakura C, Mizutani M. Hereditary hypotrophic axonopathy with neurofilament deficiency in a mutant strain of the Japanese quail. Acta Neuropathol (Berl) 1991; 82(6):427-434.
31. Xu Z, Marszalek JR, Lee MK et al. Subunit composition of neurofilaments specifies axonal diameter. J Cell Biol 1996; 133(5):1061-1069.
32. Zhu Q, Couillard-Despres S, Julien JP. Delayed maturation of regenerating myelinated axons in mice lacking neurofilaments. Exp Neurol 1997; 148(1):299-316.
33. Sternberger LA, Sternberger NH. Monoclonal antibodies distinguish phosphorylated and nonphosphorylated forms of neurofilaments in situ. Proc Natl Acad Sci USA 1983; 80(19):6126-6130.
34. Oblinger MM, Brady ST, McQuarrie IG et al. Cytotypic differences in the protein composition of the axonally transported cytoskeleton in mammalian neurons. J Neurosci 1987; 7(2):453-462.
35. Pant HC, Veeranna. Neurofilament phosphorylation. Biochem Cell Biol 1995; 73(9-10):575-592.
36. Lee VM, Otvos Jr L, Carden MJ et al. Identification of the major multiphosphorylation site in mammalian neurofilaments. Proc Natl Acad Sci USA 1988; 85(6):1998-2002.
37. Carden MJ, Goldstein ME, Bruce J et al. Studies of neurofilaments that accumulate in proximal axons of rats intoxicated with beta,beta'-iminodipropionitrile (IDPN). Neurochem Pathol 1987; 7(3):189-205.
38. Willard M, Simon C. Modulations of neurofilament axonal transport during the development of rabbit retinal ganglion cells. Cell 1983; 35(2 Pt 1):551-559.
39. Fischer I, Romano-Clarke G, Grynspan F. Calpain-mediated proteolysis of microtubule associated proteins MAP1B and MAP2 in developing brain. Neurochem Res 1991; 16(8):891-898.
40. Carden MJ, Trojanowski JQ, Schlaepfer WW et al. Two-stage expression of neurofilament polypeptides during rat neurogenesis with early establishment of adult phosphorylation patterns. J Neurosci 1987; 7(11):3489-3504.
41. Lee VM, Carden MJ, Schlaepfer WW et al. Monoclonal antibodies distinguish several differentially phosphorylated states of the two largest rat neurofilament subunits (NF-H and NF-M) and demonstrate their existence in the normal nervous system of adult rats. J Neurosci 1987; 7(11):3474-3488.
42. Hisanaga S, Ikai A, Hirokawa N. Molecular architecture of the neurofilament. I. Subunit arrangement of neurofilament L protein in the intermediate-sized filament. J Mol Biol 1990; 211(4):857-869.
43. Liem RK, Hutchison SB. Purification of individual components of the neurofilament triplet: Filament assembly from the 70 000-dalton subunit. Biochemistry 1982; 21(13):3221-3226.
44. Heins S, Wong PC, Muller S et al. The rod domain of NF-L determines neurofilament architecture, whereas the end domains specify filament assembly and network formation. J Cell Biol 1993; 123(6 Pt 1):1517-1533.
45. Ching GY, Liem RK. Assembly of type IV neuronal intermediate filaments in nonneuronal cells in the absence of preexisting cytoplasmic intermediate filaments. J Cell Biol 1993; 122(6):1323-1335.
46. Sihag RK, Nixon RA. In vivo phosphorylation of distinct domains of the 70-kilodalton neurofilament subunit involves different protein kinases. J Biol Chem 1989; 264(1):457-464.

47. Sihag RK, Nixon RA. Identification of Ser-55 as a major protein kinase A phosphorylation site on the 70-kDa subunit of neurofilaments. Early turnover during axonal transport. J Biol Chem 1991; 266(28):18861-18867.
48. Nakamura Y, Hashimoto R, Kashiwagi Y et al. Major phosphorylation site (Ser55) of neurofilament L by cyclic AMP-dependent protein kinase in rat primary neuronal culture. J Neurochem 2000; 74(3):949-959.
49. Gibb BJ, Brion JP, Brownlees J et al. Neuropathological abnormalities in transgenic mice harbouring a phosphorylation mutant neurofilament transgene. J Neurochem 1998; 70(2):492-500.
50. Hashimoto R, Nakamura Y, Goto H et al. Domain- and site-specific phosphorylation of bovine NF-L by Rho-associated kinase. Biochem Biophys Res Commun 1998; 245(2):407-411.
51. Dosemeci A, Floyd CC, Pant HC. Characterization of neurofilament-associated protein kinase activities from bovine spinal cord. Cell Mol Neurobiol 1990; 10(3):369-382.
52. Floyd CC, Grant P, Gallant PE et al. Principal neurofilament-associated protein kinase in squid axoplasm is related to casein kinase I. J Biol Chem 1991; 266(8):4987-4994.
53. Hollander BA, Bennett GS, Shaw G. Localization of sites in the tail domain of the middle molecular mass neurofilament subunit phosphorylated by a neurofilament-associated kinase and by casein kinase I. J Neurochem 1996; 66(1):412-420.
54. Link WT, Dosemeci A, Floyd CC et al. Bovine neurofilament-enriched preparations contain kinase activity similar to casein kinase I—neurofilament phosphorylation by casein kinase I (CKI). Neurosci Lett 1993; 151(1):89-93.
55. Bennett GS, Quintana R. Identification of Ser-Pro and Thr-Pro phosphorylation sites in chicken neurofilament-M tail domain. J Neurochem 1997; 68(2):534-543.
56. Shaw G, Miller R, Wang DS et al. Characterization of additional casein kinase I sites in the C-terminal "tail" region of chicken and rat neurofilament-M. J Neurochem 1997; 69(4):1729-1737.
57. Serrano L, Diaz-Nido J, Wandosell F et al. Tubulin phosphorylation by casein kinase II is similar to that found in vivo. J Cell Biol 1987; 105(4):1731-1739.
58. Serrano L, Hernandez MA, Diaz-Nido J et al. Association of casein kinase II with microtubules. Exp Cell Res 1989; 181(1):263-272.
59. Nakamura Y, Hashimoto R, Kashiwagi Y et al. Casein kinase II is responsible for phosphorylation of NF-L at Ser-473. FEBS Lett 1999; 455(1-2):83-86.
60. Yoshimura Y, Aoi C, Yamauchi T. Investigation of protein substrates of Ca(2+)/calmodulin-dependent protein kinase II translocated to the postsynaptic density. Brain Res Mol Brain Res 2000; 81(1-2):118-128.
61. Lam KY, Law SY, Chu KM et al. Gastrointestinal autonomic nerve tumor of the esophagus. A clinicopathologic, immunohistochemical, ultrastructural study of a case and review of the literature. Cancer 1996; 78(8):1651-1659.
62. Bajaj NP, al-Sarraj ST, Leigh PN et al. Cyclin dependent kinase-5 (CDK-5) phosphorylates neurofilament heavy (NF-H) chain to generate epitopes for antibodies that label neurofilament accumulations in amyotrophic lateral sclerosis (ALS) and is present in affected motor neurones in ALS. Prog Neuropsychopharmacol Biol Psychiatry 1999; 23(5):833-850.
63. Brownlees J, Yates A, Bajaj NP et al. Phosphorylation of neurofilament heavy chain side-arms by stress activated protein kinase-1b/Jun N-terminal kinase-3. J Cell Sci 2000; 113(Pt 3):401-407.
64. Giasson BI, Mushynski WE. Aberrant stress-induced phosphorylation of perikaryal neurofilaments. J Biol Chem 1996; 271(48):30404-30409.
65. Giasson BI, Mushynski WE. Study of proline-directed protein kinases involved in phosphorylation of the heavy neurofilament subunit. J Neurosci 1997; 17(24):9466-9472.
66. Guidato S, Bajaj NP, Miller CC. Cellular phosphorylation of neurofilament heavy-chain by cyclin-dependent kinase-5 masks the epitope for monoclonal antibody N52. Neurosci Lett 1996; 217(2-3):157-160.
67. Guidato S, Tsai LH, Woodgett J et al. Differential cellular phosphorylation of neurofilament heavy side-arms by glycogen synthase kinase-3 and cyclin-dependent kinase-5. J Neurochem 1996; 66(4):1698-1706.
68. Li BS, Veeranna, Grant P et al. Calcium influx and membrane depolarization induce phosphorylation of neurofilament (NF-M) KSP repeats in PC12 cells. Brain Res Mol Brain Res 1999; 70(1):84-91.
69. Li BS, Veeranna, Gu J et al. Activation of mitogen-activated protein kinases (Erk1 and Erk2) cascade results in phosphorylation of NF-M tail domains in transfected NIH 3T3 cells. Eur J Biochem 1999; 262(1):211-217.
70. Sharma M, Sharma P, Pant HC. CDK-5-mediated neurofilament phosphorylation in SHSY5Y human neuroblastoma cells. J Neurochem 1999; 73(1):79-86.

71. Roder HM, Eden PA, Ingram VM. Brain protein kinase PK40erk converts TAU into a PHF-like form as found in Alzheimer's disease. Biochem Biophys Res Commun 1993; 193(2):639-647.
72. Sun D, Leung CL, Liem RK. Phosphorylation of the high molecular weight neurofilament protein (NF-H) by Cdk5 and p35. J Biol Chem 1996; 271(24):14245-14251.
73. Veeranna, Amin ND, Ahn NG et al. Mitogen-activated protein kinases (Erk1,2) phosphorylate Lys-Ser-Pro (KSP) repeats in neurofilament proteins NF-H and NF-M. J Neurosci 1998; 18(11):4008-4021.
74. Pant AC, Veeranna, Pant HC et al. Phosphorylation of human high molecular weight neurofilament protein (hNF-H) by neuronal cyclin-dependent kinase 5 (cdk5). Brain Res 1997; 765(2):259-266.
75. Brady ST. Neurofilaments run sprints not marathons. Nat Cell Biol 2000; 2(3):E43-45.
76. Nixon RA. Dynamic behavior and organization of cytoskeletal proteins in neurons: Reconciling old and new findings. Bioessays 1998; 20(10):798-807.
77. Nixon RA. The slow axonal transport debate. Trends Cell Biol 1998; 8(3):100.
78. Nixon RA. The slow axonal transport of cytoskeletal proteins. Curr Opin Cell Biol 1998; 10(1):87-92.
79. Yabe JT, Jung C, Chan WK et al. Phospho-dependent association of neurofilament proteins with kinesin in situ. Cell Motil Cytoskeleton 2000; 45(4):249-262.
80. Yabe JT, Pimenta A, Shea TB. Kinesin-mediated transport of neurofilament protein oligomers in growing axons. J Cell Sci 1999; 112(Pt 21):3799-3814.
81. Yabe JT, Chylinski T, Wang FS et al. Neurofilaments consist of distinct populations that can be distinguished by C-terminal phosphorylation, bundling, and axonal transport rate in growing axonal neurites. J Neurosci 2001; 21(7):2195-2205.
82. Shah JV, Flanagan LA, Janmey PA et al. Bidirectional translocation of neurofilaments along microtubules mediated in part by dynein/dynactin. Mol Biol Cell 2000; 11(10):3495-3508.
83. Roy S, Coffee P, Smith G et al. Neurofilaments are transported rapidly but intermittently in axons: Implications for slow axonal transport. J Neurosci 2000; 20(18):6849-6861.
84. Wang L, Ho CL, Sun D et al. Rapid movement of axonal neurofilaments interrupted by prolonged pauses. Nat Cell Biol 2000; 2(3):137-141.
85. Yabe JT, Chan WK, Chylinski TM et al. The predominant form in which neurofilament subunits undergo axonal transport varies during axonal initiation, elongation, and maturation. Cell Motil Cytoskeleton 2001; 48(1):61-83.
86. Koehnle TJ, Brown A. Slow axonal transport of neurofilament protein in cultured neurons. J Cell Biol 1999; 144(3):447-458.
87. Jung C, Yabe JT, Lee S et al. Hypophosphorylated neurofilament subunits undergo axonal transport more rapidly than more extensively phosphorylated subunits in situ. Cell Motil Cytoskeleton 2000; 47(2):120-129.
88. Jung C, Yabe JT, Shea TB. C-terminal phosphorylation of the high molecular weight neurofilament subunit correlates with decreased neurofilament axonal transport velocity. Brain Res 2000; 856(1-2):12-19.
89. Jung C, Shea TB. Regulation of neurofilament axonal transport by phosphorylation in optic axons in situ. Cell Motil Cytoskeleton 1999; 42(3):230-240.
90. Shea TB. Microtubule motors, phosphorylation and axonal transport of neurofilaments. J Neurocytol 2000; 29(11-12):873-887.
91. Ratner N, Bloom GS, Brady ST. A role for cyclin-dependent kinase(s) in the modulation of fast anterograde axonal transport: Effects defined by olomoucine and the APC tumor suppressor protein. J Neurosci 1998; 18(19):7717-7726.
92. Strong MJ, Strong WL, Jaffe H et al. Phosphorylation state of the native high-molecular-weight neurofilament subunit protein from cervical spinal cord in sporadic amyotrophic lateral sclerosis. J Neurochem 2001; 76(5):1315-1325.
93. Julien JP, Couillard-Despres S, Meier J. Transgenic mice in the study of ALS: The role of neurofilaments. Brain Pathol 1998; 8(4):759-769.
94. Julien JP. Neurofilament functions in health and disease. Curr Opin Neurobiol 1999; 9(5):554-560.
95. Cote F, Collard JF, Julien JP. Progressive neuronopathy in transgenic mice expressing the human neurofilament heavy gene: A mouse model of amyotrophic lateral sclerosis. Cell 1993; 73(1):35-46.
96. Eyer J, Peterson A. Neurofilament-deficient axons and perikaryal aggregates in viable transgenic mice expressing a neurofilament-beta-galactosidase fusion protein. Neuron 1994; 12(2):389-405.
97. Marszalek JR, Williamson TL, Lee MK et al. Neurofilament subunit NF-H modulates axonal diameter by selectively slowing neurofilament transport. J Cell Biol 1996; 135(3):711-724.
98. Beaulieu JM, Jacomy H, Julien JP. Formation of intermediate filament protein aggregates with disparate effects in two transgenic mouse models lacking the neurofilament light subunit. J Neurosci 2000; 20(14):5321-5328.

99. Lee MK, Cleveland DW. Neurofilament function and dysfunction: Involvement in axonal growth and neuronal disease. Curr Opin Cell Biol 1994; 6(1):34-40.

100. Houseweart MK, Cleveland DW. Cytoskeletal linkers: New MAPs for old destinations. Curr Biol 1999; 9(22):R864-866.

101. Vickers JC, Morrison JH, Friedrich Jr VL et al. Age-associated and cell-type-specific neurofibrillary pathology in transgenic mice expressing the human midsized neurofilament subunit. J Neurosci 1994; 14(9):5603-5612.

102. Tu PH, Elder G, Lazzarini RA et al. Overexpression of the human NFM subunit in transgenic mice modifies the level of endogenous NFL and the phosphorylation state of NFH subunits. J Cell Biol 1995; 129(6):1629-1640.

103. Xu Z, Tung VW. Overexpression of neurofilament subunit M accelerates axonal transport of neurofilaments. Brain Res 2000; 866(1-2):326-332.

104. Elder GA, Friedrich Jr VL, Bosco P et al. Absence of the mid-sized neurofilament subunit decreases axonal calibers, levels of light neurofilament (NF-L), and neurofilament content. J Cell Biol 1998; 141(3):727-739.

105. Jacomy H, Zhu Q, Couillard-Despres S et al. Disruption of type IV intermediate filament network in mice lacking the neurofilament medium and heavy subunits. J Neurochem 1999; 73(3):972-984.

106. Meier J, Couillard-Despres S, Jacomy H et al. Extra neurofilament NF-L subunits rescue motor neuron disease caused by overexpression of the human NF-H gene in mice. J Neuropathol Exp Neurol 1999; 58(10):1099-1110.

107. Strong MJ, Sopper MM, Crow JP et al. Nitration of the low molecular weight neurofilament is equivalent in sporadic amyotrophic lateral sclerosis and control cervical spinal cord. Biochem Biophys Res Commun 1998; 248(1):157-164.

108. Williamson TL, Cleveland DW. Slowing of axonal transport is a very early event in the toxicity of ALS-linked SOD1 mutants to motor neurons. Nat Neurosci 1999; 2(1):50-56.

109. Traverse S, Gomez N, Paterson H et al. Sustained activation of the mitogen-activated protein (MAP) kinase cascade may be required for differentiation of PC12 cells. Comparison of the effects of nerve growth factor and epidermal growth factor. Biochem J 1992; 288(Pt 2):351-355.

110. Li BS, Zhang L, Gu J et al. Integrin alpha(1) beta(1)-mediated activation of cyclin-dependent kinase 5 activity is involved in neurite outgrowth and human neurofilament protein H Lys-Ser-Pro tail domain phosphorylation. J Neurosci 2000; 20(16):6055-6062.

111. Veeranna, Shetty KT, Takahashi M et al. Cdk5 and MAPK are associated with complexes of cytoskeletal proteins in rat brain. Brain Res Mol Brain Res 2000; 76(2):229-236.

112. Quarles RH. Glycoproteins of myelin sheaths. J Mol Neurosci 1997; 8(1):1-12.

113. Schachner M, Bartsch U. Multiple functions of the myelin-associated glycoprotein MAG (siglec-4a) in formation and maintenance of myelin. Glia 2000; 29(2):154-165.

114. Li C, Tropak MB, Gerlai R et al. Myelination in the absence of myelin-associated glycoprotein. Nature 1994; 369(6483):747-750.

115. Montag D, Giese KP, Bartsch U et al. Mice deficient for the myelin-associated glycoprotein show subtle abnormalities in myelin. Neuron 1994; 13(1):229-246.

116. Bartsch S, Montag D, Schachner M et al. Increased number of unmyelinated axons in optic nerves of adult mice deficient in the myelin-associated glycoprotein (MAG). Brain Res 1997; 762(1-2):231-234.

117. Li C, Trapp B, Ludwin S et al. Myelin associated glycoprotein modulates glia-axon contact in vivo. J Neurosci Res 1998; 51(2):210-217.

118. Weiss MD, Hammer J, Quarles RH. Oligodendrocytes in aging mice lacking myelin-associated glycoprotein are dystrophic but not apoptotic. J Neurosci Res 2000; 62(6):772-780.

119. Yin X, Crawford TO, Griffin JW et al. Myelin-associated glycoprotein is a myelin signal that modulates the caliber of myelinated axons. J Neurosci 1998; 18(6):1953-1962.

120. Filbin MT. The Muddle with MAG. Mol Cell Neurosci 1996; 8(2/3):84-92.

121. Song H, Ming G, He Z et al. Conversion of neuronal growth cone responses from repulsion to attraction by cyclic nucleotides. Science 1998; 281(5382):1515-1518.

122. Cai D, Shen Y, De Bellard M et al. Prior exposure to neurotrophins blocks inhibition of axonal regeneration by MAG and myelin via a cAMP-dependent mechanism. Neuron 1999; 22(1):89-101.

123. Cai D, Qiu J, Cao Z et al. Neuronal cyclic AMP controls the developmental loss in ability of axons to regenerate. J Neurosci 2001; 21(13):4731-4739.

124. Dashiell SM, Tanner SL, Pant HC et al. Myelin-associated glycoprotein modulates expression and phosphorylation of neuronal cytoskeletal elements and their associated kinases. J Neurochem 2002; 81(6):1263-1272.

CHAPTER 5

Keratin Intermediate Filaments and Diseases of the Skin

E. Birgitte Lane*

Abstract

A question that still challenges cell and tissue biologists is that of the driving forces that have selected for and conserved the numerous intermediate filament proteins in vertebrates. We are only beginning to understand the functions of this family of cyto-skeleton structures and we have very few experimental tools for testing comparative functions of these proteins in a satisfactory way. A major turning point came with the discoveries that mutations in keratin intermediate filament genes were responsible for a large number of inherited skin fragility disorders. These disease links showed unequivocally that intermediate filament proteins, at least in barrier epithelia like skin, provide essential stress resilience for cells in tissues. The keratin genes account for three quarters of all the intermediate filament genes identified in the human genome, and it seems highly likely that any function attributable these structural proteins will also be important for the nonkeratin intermediate filaments. Once the stress resistance function of intermediate filaments is taken as a fact, rather than a persistent speculation, this knowledge can guide further experimental analysis and design to allow us to at last move closer to understanding the biology of these enigmatic cytoskeleton filaments.

Epidermal Blistering Caused by Keratin Mutations

The idea that keratins may contribute a structural supporting role in epithelia shifted from hypothesis to well-accepted fact when it was discovered that mutations in keratins K5 and K14 caused profound fragility of epidermal basal keratinocytes. Analyses of a wide range of heritable skin disorders described below have confirmed that without keratin intermediate filaments to form an effective network in the cytoplasm of keratinocytes, these normally very tough epithelial cells are rendered fragile and break down when the skin is subjected to quite mild everyday physical stress, such as stretching or rubbing. As cells expressing mutant keratins break down, they destroy cell layers in the stratified epithelium.

The first of these to be identified as caused by keratin mutations was a group of disorders known as *epidermolysis bullosa simplex* (EBS), with *bullosa* referring to the bullous or fluid-filled blisters that develop on the skin following quite mild physical trauma such as scratching or rubbing. In this case the keratinocytes of the epidermal basal layer in the skin are rendered fragile by a mutation (usually dominant) in either K5 or K14,[1-3] i.e., the pair of keratins that make up the keratin cytoskeleton of these cells. The fracture of the basal cells can be seen by histology of rubbed skin and fluid accumulates in the plane of rupture. In severe cases, the

*E. Birgitte Lane—Cancer Research UK Cell Structure Research Group, Cell and Developmental Biology Division, MSI/WTB Complex, University of Dundee School of Life Sciences, Dundee DD1 5EH, U.K. Email: e.b.lane@dundee.ac.uk

Intermediate Filaments, edited by Jesus Paramio. ©2006 Landes Bioscience and Springer Science+Business Media.

Dowling-Meara type of EBS (OMIM 131760), characteristic aggregates of electron-dense material can be (variably) seen in the cytoplasm of basal cells.

The published data on mutations in keratins, and other intermediate filaments, is collected in a public internet-accessible database (www.interfil.org) maintained here at the University of Dundee. Analysis of nearly two hundred mutations in K5 and K14 to date has identified mutations in about 78% of cases diagnosed as EBS. Nearly all the identified pathogenic mutations in these genes are dominant. There are a few cases of recessive mutations which nearly all lead to loss of function, or a natural knock-out, in K14.[4-6] No K5 knock-out mutations have been identified in humans, suggesting that this is lethal, presumably because there is no "back-up" type II keratin for K5 in basal cells. (For type I keratins, there are other possible candidates to reinforce a K14 knockout, such as K15.[7])

The severest cases of EBS are mostly associated with mutation in a specific codon for an arginine at position 125 in the amino acid sequence of K14, as first described by Coulombe et al;[8] otherwise they are nearly all in the helix initiation motif or the helix termination motif, the rod ends that are so critical for intermediate filament assembly. Milder forms (so-called Weber-Cockayne type EBS (OMIM 131800), and to a certain extent the Köbner type (OMIM 131900)) are associated with mutation cluster sites outside the helix boundary peptides[9,10] but they almost certainly define other critical points in the intermediate filament proteins as the pathogenic mutations are not randomly distributed (discussed in ref. 11). Overall, the severity of a case of K5/K14 EBS depends on the position of the mutation in the keratin protein.

Candidate Genes in Other Epithelia

Once the connection between K5/K14 mutations and EBS was made, there followed a period of intense analysis of other inherited dermatological conditions whose phenotype suggested cell fragility. Amongst other benefits, this period undoubtedly brought dermatologists and basic scientists together in an interesting and highly productive way. Many more EBS mutations were found in K5 and/or K14, although as they were mostly sought only in the helix boundary regions (following the first findings) it is probable that many were missed at first. Current good practice in EBS screening must involve sequencing the whole coding region of these two genes.

It was quickly realised that other phenotypically diverse skin conditions might also be caused by mutations in (different) keratin genes. These other disorders exhibited quite distinct phenotypes from that of EBS, yet the similarities and differences of these keratin disorders all fell rapidly into place as it was seen that the diverse phenotypes depended on the tissue expression of the mutated keratin target genes. A large amount of immunohistological information on keratin expression in tissues had been gathered since monoclonal antibodies to keratins were first generated in the early 1980s, and using this information it was now possible to rapidly identify candidate genes for other epithelia fragility disorders. Mutations were soon identified by sequence analysis of these other keratin genes. Again, most transpired to be dominant negative missense mutations situated in the helix boundary peptides of keratin genes, as had been the case for K5 and K14. This enlarging group of diseases and their relationship to keratin mutations has been the subject of a number of recent reviews.[12,13] Examples of some of the mutation reports are cited below but a more extensive list of published mutations is collated in the Intermediate Filament Database (http://www.interfil.org).

The second pair of keratins in which pathogenic mutations were identified in skin fragility disorders was the epidermal secondary or differentiation-specific pair of K1 and K10, normally expressed in suprabasal post-mitotic keratinocytes as they leave the basal layer and begin their progress towards complete terminal differentiation in the epidermis. It was correctly predicted that mutations in K1 and K10 might be responsible, at least in part, for the epithelial cell breakdown seen in the epidermolytic hyperkeratosis that characterised another group of disorders, know as bullous congenital ichthyosiform erythroderma (BCIE; OMIM 113800), also sometimes called after the characteristic *epidermolytic hyperkeratosis* seen in these disorders (EHK) (see http://www.interfil.org for identified mutations). With this condition, infants can sometimes be initially misdiagnosed as EBS, but the early blistering subsides to be replaced by thick, ichthyotic skin, which can be very extensive over the body surface. Histology of such skin shows that the basal cells are intact but the suprabasal cells are fractured, which by analogy with the basal cell lysis of EBS would suggest a defect in one of the pair of suprabasal keratins, i.e., K1 or K10. This proved indeed to be the case, as several groups have confirmed with the identification of a range of dominant mutations clustering around the helix boundary motifs of either of the two proteins.[14] Mutations in K1 have also since been linked to epidermolytic palmo-plantar hyperkeratosis (EPPK; OMIM 144200), mild diffuse cases of nonepidermolytic palmoplantar keratosis (NEPPK; OMIM 600962)[15] and possibly a case of Curth-Macklin ichthyosis hystrix (OMIM 146590).[16] EPPK is more usually linked to K9 mutations;[17-19] K9 is a keratin expressed specifically in palmoplantar epidermis.

The same reasoning was applied to ichthyosis bullosa of Siemens (OMIM 146800), an uncommon epidermal fragility disorder characterised by cytolysis of upper layers of the epidermis and the formation of flat, superficial blisters. This was found to be caused by mutations in K2e,[20-22] which Franke's group had shown was a late-expressed keratin in the epidermis.[23] Thus the fragile cells again matched the known expression range of the keratin concerned.

Mutations in another pair of secondary (differentiation-specific) keratins were discovered to cause another skin fragility disorder when mutations in K4 and K13 were identified in families affected with white sponge naevus (WSN; OMIM 193900).[24-27] Loose white plaques in the buccal epithelium characterise this benign condition, which is more usually picked up by dentists than doctors. Buccal epithelium, like other regions of oro-genital epithelia, expresses K4 and K13 as the suprabasal keratin pair and WSN was found to be specifically associated with dominant mutations in these two keratins, mostly located in the helix termination motifs. Histological sections at first show a loose basketweave structure of the suprabasal cells layers, which on closer inspections can be seen to be caused by cytolysis of these cell layers.

Corneal epithelium specifically expresses keratins K3 and K12, and these two keratins have been found to harbour helix termination motif mutations in patients affected by a disorder called Meesman corneal dystrophy (MCD; OMIM 122100).[28] People suffering from MCD develop small blisters specifically on the corneal epithelium.

The most divergent phenotypes of keratin disorders lie in the pachyonychia congenita disorders. Pachyonychia (thick nails) congenita has two clinically distinct forms: the Jadassohn-Lewandowsky type (PC-1; OMIM 167200) with thick nails, oral leukoplakia and nonepidermolytic palmoplantar keratoderma (NEPPK; OMIM 600962), and the Jackson-Lawler type (PC-2; OMIM 167210) with thick nails, pilosebaceous cysts, *pili torti* (twisted or corkscrew hairs) and, variably, prematurely erupted natal teeth. Guided by candidate gene identification based on a knowledge of tissue expression patterns of keratin, PC-1 was linked with mutations in K6a or K16, and PC-2 with mutations in K6b and K17.[29-32] Milder mutations in K6b or K17 are associated with steatocystoma multiplex (OMIM 184500) suggesting that there is a severity spectrum in these genes analogous to that seen in EBS.[32,33]

There are also mutations in some hair keratin genes that are associated with hair fibre malformations such as monilethrix (OMIM 158000),[34-36] which may also be explainable by cell fragility in a critical subpopulation of cells involved in shaping the forming hair. In all these disorders the predominant location of the mutations has been found to be in the helix boundary motifs, and sometimes in the H1 domain of the head region adjacent to the N-terminal end of the rod domain. It is only in the K5/K14 mutations of EBS that a much wider range of mutation clusters along the proteins is seen. This is most likely because mutated secondary keratins are to some extent protected or reinforced by the underlying template of residual K5/K14 filaments synthesised in the basal cells and serving as an assembly template for the secondary keratins. Thus only the most disruptive type of mutations are pathogenic in the secondary keratins.

Most of the keratin disorders show skin thickening (keratoderma), especially on the palms and soles (palmoplantar skin). This is probably a result of the "chronic wounding" of cell fragmentation: cells breaking open will release cytokines to trigger a repair process, even though the epithelial barrier is not completely breached. Cell proliferation in epidermal wounding is normally part of the later phase of epidermal response, with the earlier epithelial response directed to cell migration to close the breech in the barrier. Thus the proliferative response may not need to detect free space around the cells whereas a migration response would.

From Candidate Genes to "Candidate Diseases"

For several years, there were notable omissions from the list of keratins implicated in epithelial fragility disorders, in the simple epithelial keratins such as K8/K18 (the primary keratins) and K7, K19 and K20. This might suggest that simple epithelial keratins were not so essential for cell resilience and survival as they mostly occur in internal epithelia that are not so greatly stressed as epidermal tissues. On the other hand, K8/K18 are expressed in the very earliest stages of embryogenesis, and embryos of homozygous knockout mice have greatly reduced viability.[37] The tissue expression patterns of K8/K18 suggested to us that intestinal epithelia might be vulnerable to K8/K18 mutation-induced fragility and so we examined DNA from patients with inflammatory bowel disease. Mutations have now been identified in some cases of cryptogenic liver cirrhosis (included in OMIM 215600),[38] pancreatitis[39] and in our case from inflammatory bowel disease—both Crohn disease (OMIM 266600) and ulcerative colitis (OMIM 191390).[40] Strikingly these mutations are not in the helix boundary motifs, suggesting that mutations at such critical sites might be embryonic lethal in these early-expressed keratin genes. However neither are the mutations dominant; they occur in polygenic disorders, so at best they can only be predisposing factors. This is a good example of the necessity for lateral thinking about possible consequences of cytoskeleton defects in tissues.

Supramolecular Stability versus Molecular Flux

So how does this information sit with our prior understanding of keratin function in cells? Intermediate filaments are still far less well understood than the other cytoskeleton filamentous polymer systems of actin and tubulin. In contrast with actin and tubulin, intermediate filaments have been widely believed (somewhat erroneously) to be inert and unreactive. Intermediate filaments appear so because they form complex and often very dense meshworks of rope-like polymers in the cytoplasm that are not easily disrupted by drugs, nor easily soluble. Moreover the filaments are multi-stranded and have no net polarity, so dynamic flux is much more difficult to detect and monitor than in actin filaments and microtubules. Purified intermediate filament proteins in vitro will polymerise rapidly without any requirement for cofactors or associated proteins: if this is also the case inside cells, this only further increases the difficulty of analysing single strand kinetics.

In recent years several lines of evidence have accumulated which suggest that cytoplasmic intermediate filaments are much more dynamic than was previously thought. The increasing use of techniques such as green fluorescent protein (GFP), for example, to tag and track specific proteins in living cells has revealed that much subunit exchange takes place along filaments.[41] The speed of this flux reflects the likelihood that subunits of protein are normally flickering on and off the filaments all along their length,[42] reviewed by Helfand et al.[43] Small nonfilamentous protein particles of vimentin[44] and keratin[45] have been observed that are highly mobile and that associate with actin and tubulin polymers as well as intermediate filaments. Larger nonfilamentous aggregates accumulate in some epithelial cells during mitosis[46,47] and can be induced experimentally by intracellular injection of some antibodies.[48-50] In some diseases, nonfilamentous keratin aggregates accumulate spontaneously, such as in severe forms of the genetic skin blistering condition *epidermolysis bullosa simplex*[51] and in hepatocytes of people with alcoholic cirrhosis of the liver (Mallory bodies).[52]

Keratin aggregates in many forms are associated with a high level of phosphorylation[53-55] and elevated phosphorylation of intermediate filaments is associated with the depolymerised state, suggesting that filaments are remodelled inside cells by cycles of phosphorylation and dephosphorylation.[56-58] Specific information on phosphorylation sites and their regulating enzymes is still patchy for intermediate filaments and is best defined so far for the simple epithelial keratins K8 and K18 (see Owens and Lane[59] for a recent review).

Thus, intermediate filaments such as keratins clearly have mechanisms in place that can facilitate rapid spatial modulation at many levels, from subunit flux along the filament surface to total filament collapse and remodelling by phosphorylation. Their apparent stability is probably due to the multistranded structure of the filament bundles seen by light microscopy which protects these cytoskeleton structures from total catastrophic collapse as can take place in microtubules and actin filaments. Thus, under most circumstances the tonofilament as a whole persists. The evolution of a system with this supramolecular stability indicates that network persistence is an important aspect of intermediate filament function. Finally, the recent observation that accumulation of keratin aggregates in EBS-derived keratinocytes is exacerbated by mechanical stress[60] indicates that keratin filament remodelling can take place as a direct response to mechanical stress, thus putting all the pieces in place for keratins to function as mechanical stress sensors in epithelia.

Keratin Filaments as Tensile Structures

Keratins are the intermediate filament proteins characteristic of epithelial sheet tissues and account for up to 80% of the total cell protein in differentiated keratinocytes. A typical immunofluorescence image of keratin filaments (Fig. 1) shows a dense mesh of cytoplasmic fibres running in all directions through the cell in bundles of varying thickness, linked into desmosomes at cell to cell junctions and to hemidesmosomes at the cell-substrate interface. Keratin staining images from cells in culture and in tissue sections all show evidence of trans-tissue continuity of keratin filament bundles (tonofilaments) via cell junctions.

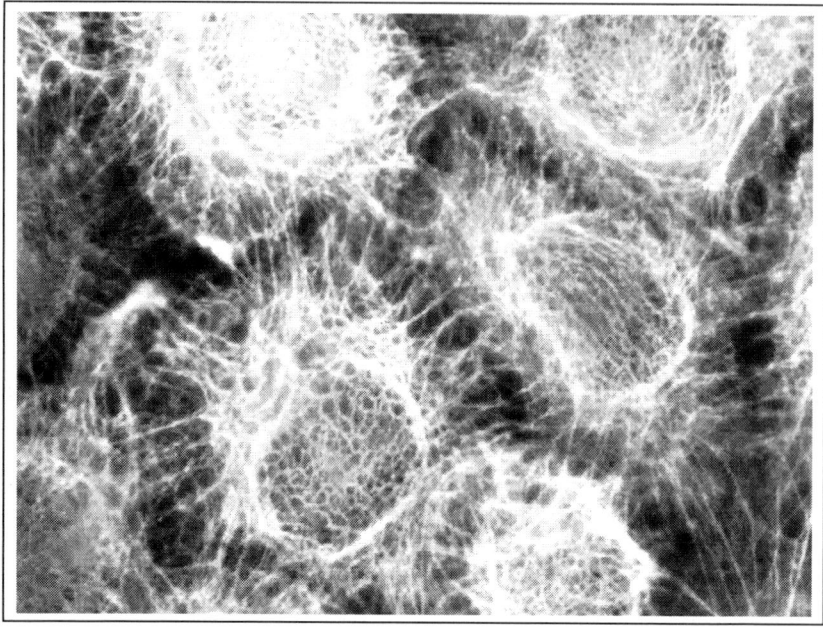

Figure 1. Immunofluorescence image of confluent keratinocytes stained with antibody to keratin, illustrating the radial tonofilaments that extend from the cell periphery into the cell body. They are typically in direct alignment with filaments subtending the other half of the desmosome in the neighbouring cell, indicating tension in the system. Deeper into the same cell the filament arrays are typically less straight and tend to track around the nucleus, suggesting lower filament tension exists simultaneously in other parts of a cell. (NEB-1 cells cultured on tissue culturegrade plastic dishes and stained with polyclonal rabbit antiserum to the C-terminal peptide of K5. Scale bar = 10 μm. Micrograph courtesy of Alison Hill.)

Do these keratin filaments sustain tension? The old term "tonofilaments" for keratin filament bundles presumes a degree of tension in the system, probably suggested by the straight arrangement of peripheral radial keratin bundles that subtend the desmosomes in keratinocytes (see Fig. 1). This first impression probably merits closer inspection as if there is tension within the keratin cytoskeleton of an epithelial cell, it is probably not evenly distributed. Seen with fluorescence light microscopy, the peripheral radial filament bundles appear run end-on into desmosomes (although by electron microscopy they appear to loop through the plaques) and are in directional alignment with their reciprocal fibres on the other side of the desmosomes in the neighbouring linked cell, suggesting a pull on both sides of a desmosome.

Further back into the cell body, these straight tonofilaments mostly become subsumed into a more ring-shaped arrangement of filaments that encircle the nucleus in the plane of the substrate. This suggests that any tension on the desmosomes must now have been dissipated or redistributed between a different set of vectors. Time-lapse filming of GFP-keratins in living cells shows sinuous wave-like flexing of keratin bundles,[41] indicating a reduced tension in at least some parts of the system. Electron microscopic images suggest that the filaments do not end at the junctions but loop through the desmosomes/hemidesmosome plaques.[61] This would intuitively seem to be functionally important, as it suggests side-on associations of filaments with junction proteins, i.e., associations involving sites that repeat along the filaments. This would allow slippage of filaments through or past the anchorage junctions and so preventing rupture of structures by mechanical stress in the epithelial sheet. Thus the cytoplasmic distribution of keratin tonofilaments suggests that some locally variable tension is maintained in the system, and the evidence for dynamic turnover suggests that filaments can be repositioned, possibly in response to mechanical cues, providing the basis for a mechanosensory system based on intermediate filaments.

Significance of Keratin Role in Skin Fragility Disorders

Disease associations have now been identified for 15 of the 21 well-studied keratins expressed in soft epithelial cells. Those not so far associated with pathogenic mutations are K2p, K7, K15, K19 and K20. K15 seems to be a secondary basal cell keratin in keratinocytes, associated with a long-term stability state, and K7, K19 and K20 are all secondary simple epithelial keratins. At least 4 other nonkeratin intermediate filaments have been associated with diverse pathologies from premature ageing (lamin mutations: Hutchinson-Gilford progeria (OMIM 176670), Emery-Dreifuss muscular dystrophy (OMIM 181350) and familial partial lipodystrophy, OMIM 151660), to desmin mutations in myofibrillar myopathy (OMIM 601419) and GFAP mutations in Alexander disease (OMIM 203450) and neurofilament mutations in some cases of ALS (OMIM 105400) (reviewed in Lane and Pekny, 2004).

The hypothesis that keratins have a role in maintaining mechanical resilience in epithelia arises very naturally from observing the configuration of keratin filament bundles (tonofilaments) and desmosome cell-cell junctions by immunofluorescence microscopy. Until the discovery of keratin mutations in skin disorders, there had been no way of proving it. Keratins are hard to get rid of: in cultured cells they appear to be refractory to most of the drugs used to disrupt actin and tubulin systems, and even microinjection of antibodies, which can be very effective in specifically disrupting keratins as we showed earlier,[48,62] did not destroy epithelial cells in tissue culture. This is probably because there is no significant mechanical stress involved in cells attached and spread on the stiff plastic substrate of a tissue culture dish.

The keratin mutations clearly show that cell fragility arises from loss of cell resilience , due to loss of function of the keratin cytoskeleton. With the filament system compromised by certain types of mutations in one of the relevant keratin genes, the cell's internal reinforcement is inadequate and the cell is torn apart by mechanical forces acting on the skin that would normally cause no detectable response in the tissue. The mechanisms for the nonkeratin disorders seem more obscure at first sight. However it seems likely that these too will eventually be recognized as resulting from structural fragility. More intermediate filament links with diseases will continue to slowly emerge as we become better at predicting the consequences of intermediate filament mutations and as large-scale sequencing becomes more trivial.

Therefore, besides the not-insignificant benefit of being able to inform patients about their disease, and to carry out prenatal screening where appropriate and where required, what else can we learn from these disorders? We should start to consider the downstream consequences of the findings in human skin diseases.

First, the disease associations show beyond doubt that keratin intermediate filament proteins are required to provide essential resilience to cells. Thus this concept of intermediate filament function has now progressed from the status of a persistent but logical speculation to that of a text-book fact. Armed with this fact, we can begin to progress to the next paradigm level of understanding intermediate filaments, such as querying the biological need for different keratins in different tissues, and from there to probe the likely evolutionary benefit of so many different intermediate filament genes. The simplest explanation for why vertebrates need different stress-resisting intermediate filaments in different tissues is that the filament proteins are subtly different and that the fine-tuning of the filament cytoskeleton in different tissues has evolved to match different tissue cell requirements.

Secondly, if it is accepted that different intermediate filaments impart different resilience characteristics to cells of different tissues, and that this stress resilience is essential for tissue function, then it is likely that the ability to sense mechanical stress and initiate mechanical stress-specific signal cascades in tissues, especially epithelia, is also an important feature of all tissues expressing specific intermediate filaments. This is a concept that has received very little attention to date except in specialised tissues like muscle. It is now clear that keratins and other intermediate filaments have mechanisms in the cell for rapid remodelling, and that this remodelling can be triggered by mechanical stress such as cell deformation by stretching[60] or swelling;[63] stress remodelling of intermediate filament has been seen in other cell types too.[64,65] All the mechanisms are in place for intermediate filaments to function as stress sensors in tissue cells; the signal transduction pathways triggered by biological tissue deformation events now need to be identified.

References

1. Bonifas JM, Rothman AL, Epstein Jr EH. Epidermolysis bullosa simplex: Evidence in two families for keratin gene abnormalities. Science 1991; 254(5035):1202-1205.
2. Vassar R, Coulombe PA, Degenstein L et al. Mutant keratin expression in transgenic mice causes marked abnormalities resembling a human genetic skin disease. Cell 1991; 64(2):365-380.
3. Lane EB, Rugg EL, Navsaria H et al. A mutation in the conserved helix termination peptide of keratin 5 in hereditary skin blistering. Nature 1992; 356(6366):244-246.
4. Rugg EL, McLean WH, Lane EB et al. A functional "knockout" of human keratin 14. Genes Dev 1994; 8(21):2563-2573.
5. Chan Y, Anton-Lamprecht I, Yu QC et al. A human keratin 14 "knockout": The absence of K14 leads to severe epidermolysis bullosa simplex and a function for an intermediate filament protein. Genes Dev 1994; 8(21):2574-2587.
6. Jonkman MF, Heeres K, Pas HH et al. Effects of keratin 14 ablation on the clinical and cellular phenotype in a kindred with recessive epidermolysis bullosa simplex. J Invest Dermatol 1996; 107(5):764-769.
7. Lloyd C, Yu QC, Cheng J et al. The basal keratin network of stratified squamous epithelia: Defining K15 function in the absence of K14. J Cell Biol 1995; 129(5):1329-1344.
8. Coulombe PA, Hutton ME, Letai A et al. Point mutations in human keratin 14 genes of epidermolysis bullosa simplex patients: Genetic and functional analyses. Cell 1991; 66(6):1301-1311.
9. Letai A, Coulombe PA, McCormick MB et al. Disease severity correlates with position of keratin point mutations in patients with epidermolysis bullosa simplex. Proc Natl Acad Sci USA 1993; 90(8):3197-3201.
10. Rugg EL, Morley SM, Smith FJ et al. Missing links: Weber-Cockayne keratin mutations implicate the L12 linker domain in effective cytoskeleton function. Nat Genet 1993; 5(3):294-300.
11. Lane EB. Keratin diseases. Curr Opin Genet Dev 1994; 4(3):412-418.

12. Smith F. The molecular genetics of keratin disorders. Am J Clin Dermatol 2003; 4(5):347-364.
13. Lane EB, McLean WHI. Keratins and skin disorders. J Pathol 2004; 204(4):355-366.
14. Rothnagel JA, Fisher MP, Axtell SM et al. A mutational hot spot in keratin 10 (KRT 10) in patients with epidermolytic hyperkeratosis. Hum Mol Genet 1993; 2(12):2147-2150.
15. Kimonis V, DiGiovanna JJ, Yang JM et al. A mutation in the V1 end domain of keratin 1 in nonepidermolytic palmar-plantar keratoderma. J Invest Dermatol 1994; 103(6):764-769.
16. Sprecher E, Ishida-Yamamoto A, Becker OM et al. Evidence for novel functions of the keratin tail emerging from a mutation causing ichthyosis hystrix. J Invest Dermatol 2001; 116(4):511-519.
17. Bonifas JM, Matsumura K, Chen MA et al. Mutations of keratin 9 in two families with palmoplantar epidermolytic hyperkeratosis. J Invest Dermatol 1994; 103(4):4740-4477.
18. Navsaria HA, Swensson O, Ratnavel RC et al. Ultrastructural changes resulting from keratin-9 gene mutations in two families with epidermolytic palmoplantar keratoderma. J Invest Dermatol 1995; 104(3):425-429.
19. Rothnagel JA, Wojcik S, Liefer KM et al. Mutations in the 1A domain of keratin 9 in patients with epidermolytic palmoplantar keratoderma. J Invest Dermatol 1995; 104(3):430-433.
20. Kremer H, Zeeuwen P, McLean WH et al. Ichthyosis bullosa of Siemens is caused by mutations in the keratin 2e gene. J Invest Dermatol 1994; 103(3):286-289.
21. Rothnagel JA, Traupe H, Wojcik S et al. Mutations in the rod domain of keratin 2e in patients with ichthyosis bullosa of Siemens. Nat Genet 1994; 7(4):485-490.
22. Arin MJ, Longley MA, Epstein Jr EH et al. A novel mutation in the 1A domain of keratin 2e in ichthyosis bullosa of Siemens. J Invest Dermatol 1999; 112(3):380-382.
23. Collin C, Moll R, Kubicka S et al. Characterization of human cytokeratin 2, an epidermal cytoskeletal protein synthesized late during differentiation. Exp Cell Res 1992; 202(1):132-141.
24. Rugg EL, McLean WH, Allison WE et al. A mutation in the mucosal keratin K4 is associated with oral white sponge nevus. Nat Genet 1995; 11(4):450-452.
25. Richard G, DeLaurenzi V, Didona B et al. Keratin-13 point mutation underlies the hereditary mucosal epithelia disorder white sponge nevus. Nature Genet 1995; 11(4):453-455.
26. Terrinoni A, Rugg EL, Lane EB et al. A novel mutation in the keratin 13 gene causing oral white sponge nevus. J Dent Res 2001; 80(3):919-923.
27. Shibuya Y, Zhang J, Yokoo S et al. Constitutional mutation of keratin 13 gene in familial white sponge nevus. Oral Surg Oral Med Oral Pathol Oral Radiol Endod 2003; 96(5):561-565.
28. Corden LD, Swensson O, Swensson B et al. Molecular genetics of Meesmann's corneal dystrophy: Ancestral and novel mutations in keratin 12 (K12) and complete sequence of the human KRT12 gene. Exp Eye Res 2000; 70(1):41-49.
29. McLean WH, Rugg EL, Lunny DP et al. Keratin 16 and keratin 17 mutations cause pachyonychia congenita. Nat Genet 1995; 9(3):273-278.
30. Bowden PE, Haley JL, Kansky A et al. Mutation of a type II keratin gene (K6a) in pachyonychia congenita. Nat Genet 1995; 10(3):363-365.
31. Smith FJ, Corden LD, Rugg EL et al. Missense mutations in keratin 17 cause either pachyonychia congenita type 2 or a phenotype resembling steatocystoma multiplex. J Invest Dermatol 1997; 108(2):220-223.
32. Smith FJ, Jonkman MF, van Goor H et al. A mutation in human keratin K6b produces a phenocopy of the K17 disorder pachyonychia congenita type 2. Hum Mol Genet 1998; 7(7):1143-1148.
33. Covello SP, Smith FJ, Sillevis Smitt JH et al. Keratin 17 mutations cause either steatocystoma multiplex or pachyonychia congenita type 2. Br J Dermatol 1998; 139(3):4750-480.
34. Winter H, Rogers MA, Gebhardt M et al. A new mutation in the type II hair cortex keratin hHb1 involved in the inherited hair disorder monilethrix. Hum Genet 1997; 101(2):165-169.
35. Winter H, Labreze C, Chapalain V et al. A variable monilethrix phenotype associated with a novel mutation, Glu402Lys, in the helix termination motif of the type II hair keratin hHb1. J Invest Dermatol 1998; 111(1):169-172.
36. Winter H, Vabres P, Larregue M et al. A novel missense mutation, A118E, in the helix initiation motif of the type II hair cortex keratin hHb6, causing monilethrix. Hum Hered 2000; 50(5):322-324.
37. Baribault H, Price J, Miyai K et al. Mid-gestational lethality in mice lacking keratin 8. Genes Dev 1993; 7(7A):1191-1202.
38. Ku NO, Wright TL, Terrault NA et al. Mutation of human keratin 18 in association with cryptogenic cirrhosis. J Clin Invest 1997; 99(1):19-23.
39. Cavestro GM, Frulloni L, Nouvenne A et al. Association of keratin 8 gene mutation with chronic pancreatitis. Dig Liver Dis 2003; 35(6):416-420.

40. Owens DW, Wilson NJ, Hill AJ et al. Human keratin 8 mutations that disturb filament assembly observed in inflammatory bowel disease patients. J Cell Sci 2004; 117(Pt 10):1989-1999.

41. Yoon KH, Yoon M, Moir RD et al. Insights into the dynamic properties of keratin intermediate filaments in living epithelial cells. J Cell Biol 2001; 153(3):503-516.

42. Vikstrom KL, Lim SS, Goldman RD et al. Steady state dynamics of intermediate filament networks. J Cell Biol 1992; 118(1):121-129.

43. Helfand BT, Chang L, Goldman RD. The dynamic and motile properties of intermediate filaments. Annu Rev Cell Dev Biol 2003; 19:445-467.

44. Prahlad V, Yoon M, Moir RD et al. Rapid movements of vimentin on microtubule tracks: Kinesin-dependent assembly of intermediate filament networks. J Cell Biol 1998; 143(1):159-170.

45. Liovic M, Mogensen MM, Prescott AR et al. Observation of keratin particles showing fast bidirectional movement colocalized with microtubules. J Cell Sci 2003; 116(Pt 8):1417-1427.

46. Horwitz B, Kupfer H, Eshhar Z et al. Reorganization of arrays of prekeratin filaments during mitosis. Immunofluorescence microscopy with multiclonal and monoclonal prekeratin antibodies. Exp Cell Res 1981; 134(2):281-290.

47. Lane EB, Goodman SL, Trejdosiewicz LK. Disruption of the keratin filament network during epithelial cell division. EMBO J 1982; 1(11):1365-1372.

48. Klymkowsky MW, Miller RH, Lane EB. Morphology, behavior, and interaction of cultured epithelial cells after the antibody-induced disruption of keratin filament organization. J Cell Biol 1983; 96(2):494-509.

49. Tolle HG, Weber K, Osborn M. Microinjection of monoclonal antibodies specific for one intermediate filament protein in cells containing multiple keratins allow insight into the composition of particular 10 nm filaments. Eur J Cell Biol 1985; 38(2):234-244.

50. Waseem A, Karsten U, Leigh IM et al. Conformational changes in the rod domain of human keratin 8 following heterotypic association with keratin 18 and its implication for filament stability. Biochemistry 2004; 43(5):1283-1295.

51. Ishida-Yamamoto A, McGrath JA, Chapman SJ et al. Epidermolysis bullosa simplex (Dowling-Meara type) is a genetic disease characterized by an abnormal keratin-filament network involving keratins K5 and K14. J Invest Dermatol 1991; 97(6):959-968.

52. Stumptner C, Omary MB, Fickert P et al. Hepatocyte cytokeratins are hyperphosphorylated at multiple sites in human alcoholic hepatitis and in a mallory body mouse model. Am J Pathol 2000; 156(1):77-90.

53. Chou CF, Omary MB. Mitotic arrest with anti-microtubule agents or okadaic acid is associated with increased glycoprotein terminal GlcNAc's. J Cell Sci 1994; 107(Pt 7):1833-1843.

54. Blankson H, Holen I, Seglen PO. Disruption of the cytokeratin cytoskeleton and inhibition of hepatocytic autophagy by okadaic acid. Exp Cell Res 1995; 218(2):522-530.

55. Strnad P, Windoffer R, Leube RE. Induction of rapid and reversible cytokeratin filament network remodeling by inhibition of tyrosine phosphatases. J Cell Sci 2002; 115(Pt 21):4133-4148.

56. Ando S, Tanabe K, Gonda Y et al. Domain- and sequence-specific phosphorylation of vimentin induces disassembly of the filament structure. Biochemistry 1989; 28(7):2974-2979.

57. Inagaki M, Gonda Y, Matsuyama M et al. Intermediate filament reconstitution in vitro. The role of phosphorylation on the assembly-disassembly of desmin. J Biol Chem 1988; 263(12):5970-5978.

58. Liao J, Ku NO, Omary MB. Stress, apoptosis, and mitosis induce phosphorylation of human keratin 8 at Ser-73 in tissues and cultured cells. J Biol Chem 1997; 272(28):17565-17573.

59. Owens DW, Lane EB. The quest for the function of simple epithelial keratins. Bioessays 2003; 25(8):748-758.

60. Russell D, Andrews PD, James J et al. Mechanical stress induces profound remodelling of keratin filaments and cell junctions in epidermolysis bullosa simplex keratinocytes. J Cell Sci 2004; 117(Pt 22):5233-5243.

61. Fawcett DW. The Cell. 2nd ed. WB. Saunders, 1981.

62. Lane EB, Klymkowsky MW. Epithelial tonofilaments: Investigating their form and function using monoclonal antibodies. Cold Spring Harb Symp Quant Biol 1982; 46(Pt 1):387-402.

63. D'Alessandro M, Russell D, Morley SM et al. Keratin mutations of epidermolysis bullosa simplex alter the kinetics of stress response to osmotic shock. J Cell Sci 2002; 115(Pt 22):4341-4351.

64. Helmke BP, Goldman RD, Davies PF. Rapid displacement of vimentin intermediate filaments in living endothelial cells exposed to flow. Circ Res 2000; 86(7):745-752.

65. Helmke BP, Thakker DB, Goldman RD et al. Spatiotemporal analysis of flow-induced intermediate filament displacement in living endothelial cells. Biophys J 2001; 80(1):184-194.

CHAPTER 6

The Keratin K6 Minifamily of Genes

Manuel Navarro*

Abstract

Keratin K6 constitutes a special case among the keratin intermediate filaments. It is constitutively expressed in several stratified epithelia, but is also induced by several stimuli, many of which are related to hyperproliferation. In addition, this keratin is, unlike others, encoded by several genes, which give rise to similar but not identical forms. In recent years, considerable advances have been made in the identification of new K6 genes and the understanding of the function of K6. Here I review the present knowledge about the human, murine and bovine keratin K6 genes, in particular with regard to the differences in sequence among the different isoforms and their different regulation. Hints about the possible K6 biological function that are suggested by the study of murine models of overexpression and gene inactivation, as well as by the study of human diseases due to mutations in K6, are also discussed.

Introduction

Keratins are the most abundant structural proteins in epithelial cells, where they constitute the cytoskeleton of intermediate filaments. In fact, skin epithelial cells are so loaded with keratins that they were initially denominated keratinocytes. Consistent with this abundance of proteins, keratin genes are among the most intensely expressed in keratinocytes, both in vivo and in vitro (Sesto and Navarro, unpublished results). Keratins constitute the most extensive protein family among the intermediate filaments: in humans, some 30 different keratin polypeptides have been described, which can be clustered in two different types, I and II, based on their sequence and genomic structure.[1] For establishing their structural network, keratins must assemble in tetramers, which consist of specific pairs of type I and type II keratins. Expression of these pairs occurs in a tissue- and differentiation-specific mode, in such a way that the different epithelial cell types can be characterized by the keratin pairs they express.[2,3] It can be easily seen that the expression of keratins must be tightly regulated to obtain stechiometry and cell type specificity.

One of the most interesting members of the keratin family is K6 which, together with its partners K16 and K17, is constitutively expressed in a series of internal stratified epithelia such as esophagus, tongue, palate, and female genital tract.[2,4,5] In the skin and adnexa, it is expressed in the suprabasal layers of palmar and plantar epidermis, nails,[4] and in the hair follicle outer root sheath.[6] What makes K6 extraordinary is that, in addition to having (as every other keratin) the "standard" constitutive expression in its own characteristic array of tissues, it can also be induced in a vast number of skin hyperproliferative situations, such as skin carcinomas, psoriasis, warts, inflammatory processes, and wound healing.[7,8] K6 is also induced in cell culture,[9] and responds to several chemicals with proliferative effects, such as retinoic acid or TPA.[10-12] Thus, the expression of K6 goes beyond the characteristic pattern of expression of

*Manuel Navarro—Department of Molecular and Cell Biology, CIEMAT, Madrid, Spain.
Email: manuel.navarro@ciemat.es

Intermediate Filaments, edited by Jesus Paramio. ©2006 Landes Bioscience
and Springer Science+Business Media.

other keratins, existing a number of situations in which K6 seems to be expressed upon demand. In fact, it has been proposed that K6 induction occurs whenever the normal biology of keratinocytes is disturbed, irrespective of their proliferative status.[13]

K6 is also unusual in that it is coded by several genes, which are transcribed to very similar (but no identical) proteins. Initial studies described two genes for human K6,[14] and later it was discovered that at least six different genes encode this keratin in our species.[15] Recently, new K6 related genes have been identified during a screening of the draft sequence of the human genome.[16] Since some of the formerly known genes were not found in the draft sequence of the human genome, it is still possible that additional K6 genes appear. In mice, two different K6 genes have been described so far,[17,18] and three in bovines.[12,19,20] In addition to all these genes, there are at least four human[21-23] and one murine[23,24] K6-related K6irs keratins, which are specifically expressed in different cells of the inner root sheath of the hair follicle. Interestingly, K6irs1 and K6irs2 sequences coincide, respectively, with the hK6i and hK6k genes identified in the analysis of the human genome draft sequence,[16] but the sequences of K6irs3 and K6irs4 were not identified in that study. A human K6hf has also been identified which is specifically expressed in the innermost (or companion) layer of the outer root sheath of the hair follicle.[25] Similarly, a murine mK6hf has also been described.[26] The type I partner for all these keratins have not yet been identified.[25] The sequences of some of these keratins are most closely related to other keratins than K6 (for instance, hK6hf is more similar to hK5, and hK6irs1 to hK4)[22,24,25] but, since they comigrate with K6 in one-dimensional SDS protein gels, and are expressed in the hair follicle, they have been designated as "K6", according to the original keratin classification criteria.[2,25] However, evolutionary tree analysis of these keratins locates them away from the "bona fide" K6 genes, and in fact the four different human K6irs represent a different subgroup of keratins, as hair keratins do.[21]

Pseudogenes have been found among human keratins, but most of them belong to the K8 and K18 genes.[16] No pseudogene has been found, up to date, for human K6 genes, except one for hK6hf,[16] the keratin specific of the companion layer of the hair follicle.[25] In mice, however, two K6 pseudogenes have been described.[18] No pseudogenes have been reported for bovine K6.

Why are there some many K6 proteins? The multiplicity of K6 genes in several species strongly suggests that this represents an advantage for the organism. This idea is reinforced by the evolutive tree analysis of K6, which suggests that the different genes evolved independently after species separated.[18] It is possible that the different K6 polypeptides could have different regulation and/or different functions. They might respond to different stimuli or be gradually induced in such a way that the keratinocyte can get the needed amount of K6 in a fast way.

Conservation of Sequence in the K6 Polypeptides

Keratin sequences are highly conserved in their central, coiled-coil region, while the regions in 5' (head) and 3' (tail) are specific to each keratin. In the last years, it has become evident that the different keratins have different functional properties, and it has been shown that, while the central, conserved region is used for assembly, the amino- and carboxi-terminal domains confer to keratins their different functional properties.[27-29] A study of the 5' and 3' sequences of the different K6 polypeptides could shed some light on their possible differential properties. Comparison of the amino acid sequences of the known human, murine and bovine K6 genes reveals, as expected, a very high identity in the available α-helical region (not shown) but also an elevated identity in the amino terminal region (Fig. 1). It can be observed that keratins of the inner root sheath, both human and murine, clearly constitute a separate group, as happens with hK6hf and mK6hf. These K6hf keratins, however, are not as different from the "bona fide" K6 sequences as the K6irs are. hK6l also diverges from all the other K6 sequences, but has strong sequence homology with the protein encoded by the murine cDNA BC031593[30] (comparison not shown), which might represent an additional, not previously identified mouse K6 gene. On the basis of this comparison, it is predicted that hK6l and BC031593 constitute a new group of K6-related keratins.

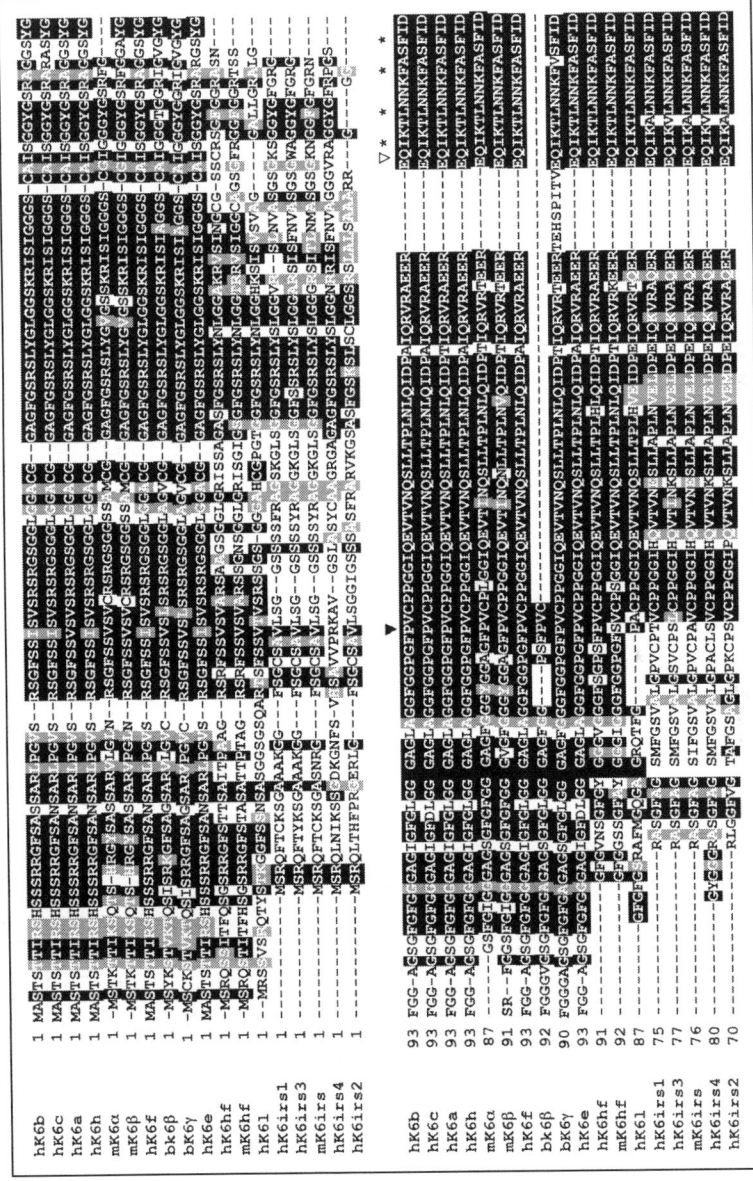

Figure 1. Comparison of the 5'-aminoterminal domains of the human, murine and bovine K6 proteins so far identified. Note that the sequences of hK6d and bK6α in this region are not available, and the sequence of bK6β is truncated in 3'. Boxes shaded in black represent residues that are identical in at least 50% of the proteins; boxes shaded in gray indicate residues with similar properties. Sequences were aligned with ClustalW 1.8 and shaded with Boxshade. The beginnings of the H1 and the coiled-coil domains are indicated (filled and open arrowheads, respectively), as are the heptad repeats of hydrophobic residues (asterisks).

Considered by species, the human hK6a, hK6b, hK6c, hK6e, hK6f and hK6h polypeptides share considerable sequence homology. Keratins hK6h and hK6f are identical in this 5' aminoterminal region, and among the other keratin polypeptides, only six residues in the tract of 163 vary,[15] with some of these differences resulting in aminoacids with similar properties (e.g., valine-isoleucine), and others that affect to phosphorilable residues. Study of the putative molecules that could bind to these variable residues[31] allows us to see that in hK6c and hK6e the presence of an arginine instead of a glycine in position 88 confers a greater surface accessibility and gives rise to a putative PKCδ phosphorilation site in the adjacent serine 90. Interestingly, PKCδ serves as receptor for phorbol esters. Similarly, the presence of a serine in position 21 of hK6b creates a putative phosphorilation site for clk2 kinase. Thus, the possibility exists that these small sequence differences between the human K6 polypeptides may translate into differential binding to other molecules, which may result in different functional capabilities for the different K6 isoforms. However, these statements should be treated with caution, since they are based only in "in silico" analysis, and up to the date there is no proof of functional differences among the several K6 keratins.

In the H1 domain of all K6 keratins the sequence LLTPL, which is a target for phosphorilation related to stress situations,[32] is conserved, as is in all type II keratins. Notably, in the K6irs keratins, murine and human, this sequence is changed to LLAPL, where the threonine that undergoes phosphorilation is lost, and so these K6irs are predicted not to be sensitive to stress-induced depolymerization.

In the case of the murine polypeptides, the differences between the two isoforms consist mainly in the deletion of an eight-aminoacid tract in mK6α. In the bovine genes the aminoterminal sequence of bK6α is not known, but the amino acid sequences of bK6β and bK6γ are highly similar in this region (96% identity). It is seen that bK6γ shows more similarity to the human sequences than bK6β, since in almost all positions in which bK6β and bK6γ differ, the sequence of bK6γ is identical to the human sequences. Also, the bK6β protein lacks a sequence GFGG, that is present in most of the K6 and K6hf proteins (Fig. 1). Strikingly, in bK6γ there is an eight amino acids sequence TEHSPITV that is unique among the known keratin sequences. This sequence is particularly interesting, since it is located after amino acid 165, just between the end of the H1 domain and the beginning of the alpha-helical rod domain, but its structure is not α-helical and possesses three putative phosphorilation sites. The significance of this sequence remains to be tested, but since the gene for bK6γ is expressed in tissues in which no aberrant keratin structures are found (not shown), it is possible that the only effect of this insertion is to delay the start of the α-helix. On the other hand, this sequence is located in the middle of two frequently mutated regions in keratin genes: mutations in the gene for hK6a in pachyonychia congenita concentrate in hot spots corresponding to amino acids 171 and 174,[33-36] just seven amino acids after the inclusion of the TEHSPITV sequence, and mutations in the H1 domain, which has a direct role in the filament assembly, have been found in several keratins.[37] Since the H1 sequence in bK6γ is not mutated, but only separated from the coil, and the heptads repeats of hydrophobic residues in the alpha-helix are not disrupted (see Fig. 1), it is difficult to predict how the presence of this sequence might influence the structure of keratin bK6γ or affect the stability of the proteins. Interestingly, it has been postulated that destabilization of the coil might be important for the wound healing response.[38]

Comparisons of the sequences of the carboxiterminal ends of all published human, murine and bovine keratins (Fig. 2) clearly confirm that K6irs and K6hf keratins from mice and human constitute separate groups. Here also, as suggested by the comparison of the aminoterminal domains, hK6l seems to be an "independent" keratin. Among the other keratins, the sequences here are more conserved than in the aminoterminal domain and, consistent with the idea that K6 genes evolved independently after species diverged,[18] differences between species are more evident than intraspecies, with several of the keratins being identical in this carboxiterminal region: in humans, hK6b and hK6f are identical, as are hK6a and hK6c. The hK6f and hK6c proteins, which are identical in the aminoterminal region, differ here in three

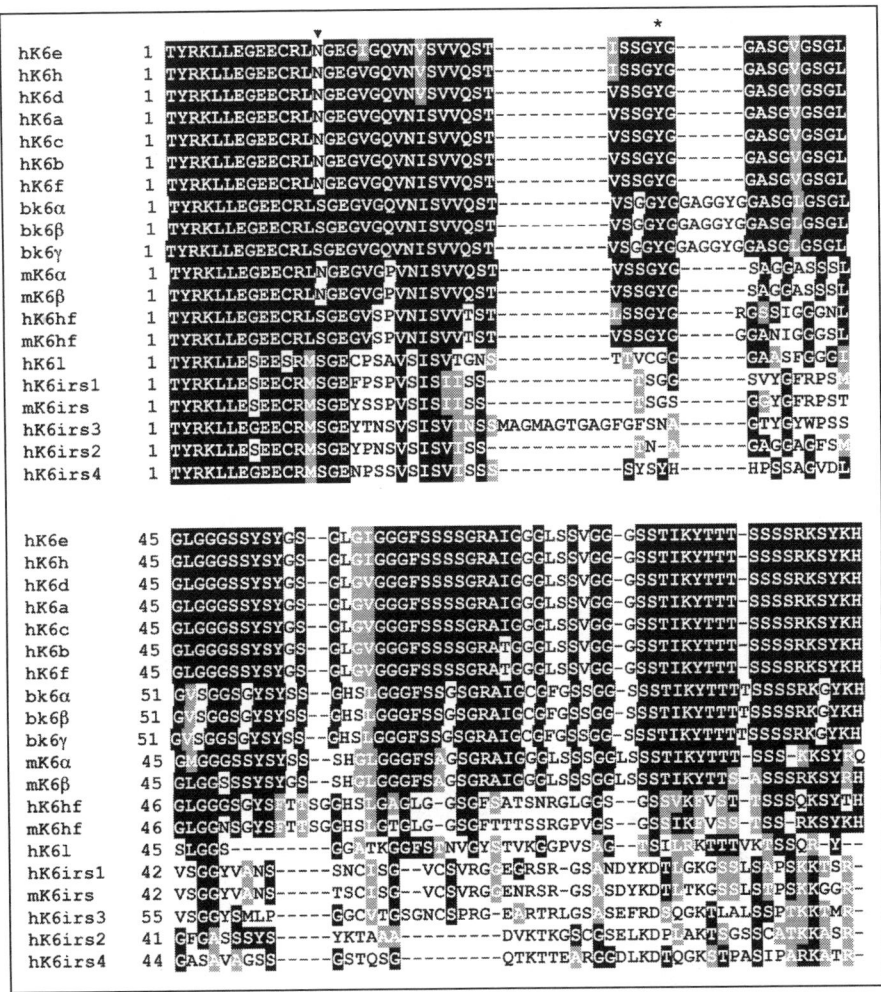

Figure 2. Comparison of the 3'-carboxiterminal domains of the human, murine and bovine K6 genes so far identified. Boxes shaded in black represent residues that are identical in at least 50% of genes; boxes shaded in gray indicate residues with similar properties. Sequences were aligned with ClustalW 1.8 and shaded with Boxshade. The ends of the coiled-coil and H2 domains are indicated (arrowhead and asterisk, respectively).

aminoacids. In bovines, the three isoforms for bK6α, bK6β and bK6γ are identical, and include an insertion GAGGYG that is not found in the human or murine proteins. A search for differences in putative protein binding did not give significant results here.

Expression of the K6 Genes

How is the regulation of the expression of this plethora of K6 genes achieved? Few studies have been performed, and maybe the main finding is that constitutive and inducible expression can be separated. Transgenic mice studies using the promoter of the bovine *bK6β* gene have elucidated the regions controlling tissue-specific and hyperproliferation-related expression of this keratin.[39,40] While the ability to induce expression resides in the 2.4 kb preceding the transcription start, the cooperation of a distal and a proximal element in a 9 kbp region is needed for the constitutive expression.[39,40] In humans, the situation could be similar, since 5.2 kb of 5'-upstream sequence of the gene coding for hK6a are not enough for conferring constitutive expression,[41] but the 960 bp preceding the transcription start suffice to confer inducible expression to a reporter gene, although an upstream enhancer is necessary for maximum inducibility.[41] Strikingly, the induction of transgenes using either the bovine *bK6β* or human *hK6a* regulatory sequences is suprabasal,[40,41] while it has been reported that the endogenous *mK6α* gene suffers mainly basal induction.[18,42]

Even less is known about the transcription factors that govern the regulation of these K6 genes. No studies have been done on the distal elements, and only short stretches of the proximal promoters of the *bK6β* and *hK6b* genes have been studied in detail. Among the functional elements characterized are AP-1, AP-2 and RARE elements in the *bK6β* gene,[12] and AP-1,[43] AP-2,[44] C/EBP[45] and an EGF-responsive sequence[46] in the *hK6b* promoter. A comparison of the 5'-upstream, noncoding sequences of the *bK6β*, *hK6a*, *hK6b*, *mK6α* and *mK6β* genes yields an elevated identity in the near 250 bp of available sequences (Fig. 3), where almost all the regulatory sequences are conserved. The AP-1 core sequence found in the *bK6β* gene[12] in the position most proximal to the TATA box is conserved in all five genes, as is the adjacent "epidermal box".[12,20] The EGF-responsive sequence characterized in the *hK6b* promoter[46] is also well conserved, although no transcription factor binding in this region was detected in the *bK6β* gene.[12] The binding sites for the transcription factor C/EBP-gadd153 found in the *hK6b*

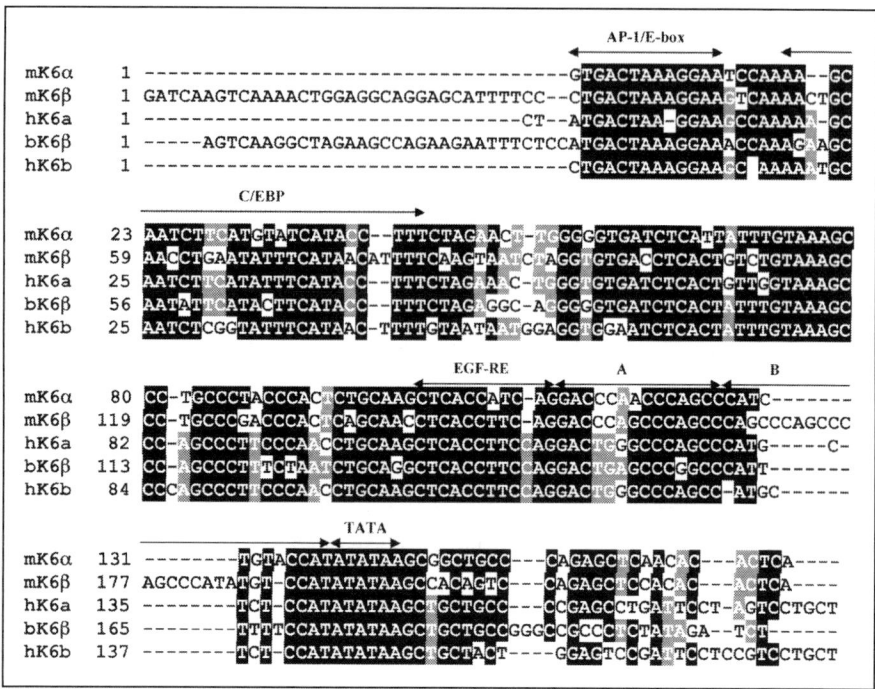

Figure 3. Comparison of the 5'-upstream sequences of (from top to bottom) *mK6α*,[18] *mK6β*,[18] *hK6a*,[15] *bK6β*[12,53] and *hK6b*.[43] The sequence comparison ends at the position of the *bK6β* capsite.[20] Boxes shaded in black indicate sequence identity in at least four of the five genes. Boxes shaded in grey represent sequence identity in three genes. The sequences that have been shown to interact with nuclear proteins in the *bK6β* 5'-upstream region (AP-1/epidermal box)[12] as well as those binding nuclear proteins in the *hK6b* gene (C/EBP,[45] EGFRE,[45,46] A and B[43]) are indicated. The TATA box is also shown. Sequences were aligned with ClustalW 1.8 and shaded with Boxshade.

promoter[45] are not so well conserved. Interestingly, a C/EBP binding site has been shown to be responsible for the TPA induction of the involucrin gene,[47] and many of the sites here described (AP-1, RARE, EGF) are related with the inducibility of the K6 genes. Going 5' upstream (not shown) the second AP-1 core sequence found in the *bK6β* gene[12] is conserved in the *hK6b* gene,[45] as is the AP-2 sequence in the *bK6β* and *hK6b* genes.[12,44]

Expression and Partners

Historically, an important question has been to elucidate if the different K6 isoforms have a similar or different pattern of expression. The elevated homology sequence between the different K6 proteins has difficulted the use of antibodies to identify their pattern of expression. In humans, some information has been obtained using probes from the less conserved 3' nontranslated regions. In this way, it has been shown that hK6a is the main isoform expressed in hair follicles, sole skin and cultured keratinocytes.[15] In addition, the study of the phenotypes of the diseases in which human K6 genes are involved (pachyonychia congenita of types I and II) has allowed to infer the pattern of expression of hK6a and hK6b, respectively: while hK6a seems to have the highest expression in most K6 positive tissues, hK6b is expressed mainly in the luminal cell layers of the sebaceous ducts.[48] It has also been determined that the partner of K6a is K16[33] and the partner of K6b is K17,[48] although sometimes K17 is found without K6b. Almost nothing is known about the expression of the other human K6 isoforms.[15] In mice, specific antibodies for mK6α and mK6β have been recently developed,[17,42] and have helped to elucidate that mK6α is the main isoform, and it is expressed mainly in the skin follicles and whiskers.[18] In tongue, palate and footpad, mK6β has been reported to be more suprabasal, while expression of mK6α apparently occurs in all layers.[42]

Functional Considerations

In spite of all the recent developments, the biology of K6 is still not well understood. Some of the K6 genes have been more or less linked to human diseases: mutations in hK6a and hK6b cause pachyonychia congenita (PC) of type 1 and 2, respectively.[34,48] PC1 (OMIM #167200) is characterized by onychogryposis, hyperkeratosis of the palms, soles, knees and elbows, tiny cutaneous horns in many areas, and leukoplakia of the oral mucous membranes. PC2 (OMIM # 167210) has natal teeth and epidermoid cysts, but no oral leukoplakia. In contrast to PC1, PC2 has minimal oral involvement and milder keratoderma, and multiple steatocystomas (cystic hamartomas lined by sebaceous ductal epithelium) is a major clinical feature.[48] Mutations in the hK6hf gene, its type I partner, or one of the K6irs genes might be responsible for the development of the loose anagen syndrome[49] (OMIM 600628). The remaining K6 polypeptides have not yet been associated to any specific disease.

The study of knock-out murine models only partially unveils the functions of K6: K6a null mice are viable and do heal wounds normally,[42] although it has been reported that they display a delayed reepithelization from the hair follicles upon mild wounding.[42] Two independent knockouts of both the murine *mK6α* and *mK6β* genes caused the early death of the animals, probably due to severe blistering in the oral mucosa.[26,50] The nails, however, did not suffer evident alterations, and apparently there were no differences in the embryonic[51] or adult[26] wound healing. As a consequence of these findings, the third murine keratin mK6hf, which is expressed in hair follicles and nails (but not in tongue, palate or footpad), was identified.[26] But these results also illustrate the difficulties in interpretation due to the different genetic background of the animals: the survival of the double K6 knockout mice was different depending on the strain.[26,50] Moreover, it has been recently shown that keratinocytes null for both mK6α and mK6β show an increased proliferative potential when put in culture, probably due to increased migration, and that these keratinocytes are destroyed when subjected to mechanical trauma,[52] but this effect does not occur in the other model of mK6α and mK6β knockout,[26] which is made in a different genetic background.

   The analysis of the function of K6 could perhaps also benefit from the separation of inducible and constitutive roles. In the case of the constitutive expression the importance of the different K6 forms is evident, as seen in the different phenotypes observed in PC1 and PC2 patients, which can be attributed to the different pattern of expression of each gene and are strong evidence of lack of compensation for the different K6 keratins. However, for the K6 induction upon wound healing the situation is a little bit more complex. Several evidences suggest that the different K6 genes have cell-type specificity of induction: in mice, mK6α seems to be responsible for the induction in the basal layer of the epidermis and the outer root sheath, while mK6β can be induced only in the suprabasal layers of the epidermis.[18,42] Induction of mK6α and mK6β occurs with different kinetics, and the mK6β gene seems to be more strongly induced.[18] On the contrary, human PC patients have not been reported to show any alteration in wound healing, and this might mean that other K6 isoform might compensate for the mutated one. However, mutations in keratin genes are usually dominant, causing a collapse in the intermediate filament cytoskeleton, which cannot be precluded by expression of other keratins. So, it is possible that the absence of these keratins is not enough to affect the wound healing response, similarly to what happens in a line of mK6α and mK6β null mice,[26] but not in other.[52] In this respect, it has been suggested that the role of K6 in wound healing may be only structural, as a support against mechanical stress.[26] K6 proteins would be able to provide the correct architecture to support both migration and resistance in the special context of cells in the margins of injured tissues.[52] If this explanation is correct, there may be other cellular factors (or perhaps K6 genes) that are able sometimes to compensate for the absence of a given isoform of K6.

References

1. Fsuchs E, Weber K. Intermediate filaments: Structure, dynamics, function, and disease. Annu Rev Biochem 1994; 63:345-382.
2. Moll R, Franke WW, Schiller DL et al. The catalog of human cytokeratins: Patterns of expression in normal epithelia, tumors and cultured cells. Cell 1982; 31(1):11-24.
3. Eichner R, Bonitz P, Sun TT. Classification of epidermal keratins according to their immunoreactivity, isoelectric point, and mode of expression. J Cell Biol 1984; 98(4):1388-1396.
4. Quinlan RA, Schiller DL, Hatzfeld M et al. Patterns of expression and organization of cytokeratin intermediate filaments. Ann NY Acad Sci 1985; 455:282-306.
5. Rentrop M, Knapp B, Winter H et al. Differential localization of distinct keratin mRNA-species in mouse tongue epithelium by in situ hybridization with specific cDNA probes. J Cell Biol 1986; 103(6 Pt 2):2583-2591.
6. Stark HJ, Breitkreutz D, Limat A et al. Keratins of the human hair follicle: "Hyperproliferative" keratins consistently expressed in outer root sheath cells in vivo and in vitro. Differentiation 1987; 35(3):236-248.
7. Stoler A, Kopan R, Duvic M et al. Use of monospecific antisera and cRNA probes to localize the major changes in keratin expression during normal and abnormal epidermal differentiation. J Cell Biol 1988; 107(2):427-446.
8. Mansbridge JN, Knapp AM. Changes in keratinocyte maturation during wound healing. J Invest Dermatol 1987; 89(3):253-263.
9. Weiss RA, Eichner R, Sun TT. Monoclonal antibody analysis of keratin expression in epidermal diseases: A 48- and 56-kdalton keratin as molecular markers for hyperproliferative keratinocytes. J Cell Biol 1984; 98(4):1397-1406.
10. Schweizer J, Furstenberger G, Winter H. Selective suppression of two postnatally acquired 70 kD and 65 kD keratin proteins during continuous treatment of adult mouse tail epidermis with vitamin A. J Invest Dermatol 1987; 89(2):125-131.
11. Molloy CJ, Laskin JD. Specific alterations in keratin biosynthesis in mouse epidermis in vivo and in explant culture following a single exposure to the tumor promoter 12-O-tetradecanoylphorbol-13-acetate. Cancer Res 1987; 47(17):4674-4680.
12. Navarro JM, Casatorres J, Jorcano JL. Elements controlling the expression and induction of the skin hyperproliferation-associated keratin K6. J Biol Chem 1995; 270(36):21362-21367.

13. Mahony D, Karunaratne S, Cam G et al. Analysis of mouse keratin 6a regulatory sequences in transgenic mice reveals constitutive, tissue-specific expression by a keratin 6a minigene. J Invest Dermatol 2000; 115(5):795-804.
14. Tyner AL, Eichman MJ, Fuchs E. The sequence of a type II keratin gene expressed in human skin: Conservation of structure among all intermediate filament genes. Proc Natl Acad Sci USA 1985; 82(14):4683-4687.
15. Takahashi K, Paladini RD, Coulombe PA. Cloning and characterization of multiple human genes and cDNAs encoding highly related type II keratin 6 isoforms. J Biol Chem 1995; 270(31):18581-18592.
16. Hesse M, Magin TM, Weber K. Genes for intermediate filament proteins and the draft sequence of the human genome: Novel keratin genes and a surprisingly high number of pseudogenes related to keratin genes 8 and 18. J Cell Sci 2001; 114(Pt 14):2569-2575.
17. Rothnagel JA, Seki T, Ogo M et al. The mouse keratin 6 isoforms are differentially expressed in the hair follicle, footpad, tongue and activated epidermis. Differentiation 1999; 65(2):119-130.
18. Takahashi K, Yan B, Yamanishi K et al. The two functional keratin 6 genes of mouse are differentially regulated and evolved independently from their human orthologs. Genomics 1998; 53(2):170-183.
19. Jorcano JL, Franz JK, Franke WW. Amino acid sequence diversity between bovine epidermal cytokeratin polypeptides of the basic (type II) subfamily as determined from cDNA clones. Differentiation 1984; 28(2):155-163.
20. Blessing M, Zentgraf H, Jorcano JL. Differentially expressed bovine cytokeratin genes. Analysis of gene linkage and evolutionary conservation of 5'-upstream sequences. EMBO J 1987; 6(3):567-575.
21. Langbein L, Rogers MA, Praetzel S et al. K6irs1, K6irs2, K6irs3, and K6irs4 represent the inner-root-sheath-specific type II epithelial keratins of the human hair follicle. J Invest Dermatol 2003; 120(4):512-522.
22. Langbein L, Rogers MA, Praetzel S et al. A novel epithelial keratin, hK6irs1, is expressed differentially in all layers of the inner root sheath, including specialized huxley cells (Flugelzellen) of the human hair follicle. J Invest Dermatol 2002; 118(5):789-799.
23. Porter RM, Corden LD, Lunny DP et al. Keratin K6irs is specific to the inner root sheath of hair follicles in mice and humans. Br J Dermatol 2001; 145(4):558-568.
24. Aoki N, Sawada S, Rogers MA et al. A novel type II cytokeratin, mK6irs, is expressed in the Huxley and Henle layers of the mouse inner root sheath. J Invest Dermatol 2001; 116(3):359-365.
25. Winter H, Langbein L, Praetzel S et al. A novel human type II cytokeratin, K6hf, specifically expressed in the companion layer of the hair follicle. J Invest Dermatol 1998; 111(6):955-962.
26. Wojcik SM, Longley MA, Roop DR. Discovery of a novel murine keratin 6 (K6) isoform explains the absence of hair and nail defects in mice deficient for K6a and K6b. J Cell Biol 2001; 154(3):619-630.
27. Paramio JM, Segrelles C, Ruiz S et al. Inhibition of protein kinase B (PKB) and PKCzeta mediates keratin K10-induced cell cycle arrest. Mol Cell Biol 2001; 21(21):7449-7459.
28. Paramio JM, Casanova ML, Segrelles C et al. Modulation of cell proliferation by cytokeratins K10 and K16. Mol Cell Biol 1999; 19(4):3086-3094.
29. Paladini RD, Coulombe PA. The functional diversity of epidermal keratins revealed by the partial rescue of the keratin 14 null phenotype by keratin 16. J Cell Biol 1999; 146(5):1185-1201.
30. Strausberg RL, Feingold EA, Grouse LH et al. Generation and initial analysis of more than 15,000 full-length human and mouse cDNA sequences. Proc Natl Acad Sci USA 2002; 99(26):16899-16903.
31. Yaffe MB, Leparc GG, Lai J et al. A motif-based profile scanning approach for genome-wide prediction of signaling pathways. Nat Biotechnol 2001; 19(4):348-353.
32. Toivola DM, Zhou Q, English LS et al. Type II keratins are phosphorylated on a unique motif during stress and mitosis in tissues and cultured cells. Mol Biol Cell 2002; 13(6):1857-1870.
33. Terrinoni A, Smith FJ, Didona B et al. Novel and recurrent mutations in the genes encoding keratins K6a, K16 and K17 in 13 cases of pachyonychia congenita. J Invest Dermatol 2001; 117(6):1391-1396.
34. Bowden PE, Haley JL, Kansky A et al. Mutation of a type II keratin gene (K6a) in pachyonychia congenita. Nat Genet 1995; 10(3):363-365.
35. Lin MT, Levy ML, Bowden PE et al. Identification of sporadic mutations in the helix initiation motif of keratin 6 in two pachyonychia congenita patients: Further evidence for a mutational hot spot. Exp Dermatol 1999; 8(2):115-119.
36. Smith FJ, McKenna KE, Irvine AD et al. A mutation detection strategy for the human keratin 6A gene and novel missense mutations in two cases of pachyonychia congenita type 1. Exp Dermatol 1999; 8(2):109-114.

37. Yang JM, Nam K, Park KB et al. A novel H1 mutation in the keratin 1 chain in epidermolytic hyperkeratosis. J Invest Dermatol 1996; 107(3):439-441.
38. Wawersik M, Paladini RD, Noensie E et al. A proline residue in the alpha-helical rod domain of type I keratin 16 destabilizes keratin heterotetramers. J Biol Chem 1997; 272(51):32557-32565.
39. Ramirez A, Vidal M, Bravo A et al. Analysis of sequences controlling tissue-specific and hyperproliferation-related keratin 6 gene expression in transgenic mice. DNA Cell Biol 1998; 17(2):177-185.
40. Ramirez A, Vidal M, Bravo A et al. A 5'-upstream region of a bovine keratin 6 gene confers tissue-specific expression and hyperproliferation-related induction in transgenic mice. Proc Natl Acad Sci USA 1995; 92(11):4783-4787.
41. Takahashi K, Coulombe PA. Defining a region of the human keratin 6a gene that confers inducible expression in stratified epithelia of transgenic mice. J Biol Chem 1997; 272(18):11979-11985.
42. Wojcik SM, Bundman DS, Roop DR. Delayed wound healing in keratin 6a knockout mice. Mol Cell Biol 2000; 20(14):5248-5255.
43. Bernerd F, Magnaldo T, Freedberg IM et al. Expression of the carcinoma-associated keratin K6 and the role of AP-1 proto-oncoproteins. Gene Expr 1993; 3(2):187-199.
44. Leask A, Byrne C, Fuchs E. Transcription factor AP2 and its role in epidermal-specific gene expression. Proc Natl Acad Sci USA 1991; 88(18):7948-7952.
45. Komine M, Rao LS, Kaneko T et al. Inflammatory versus proliferative processes in epidermis. Tumor necrosis factor alpha induces K6b keratin synthesis through a transcriptional complex containing NFkappa B and C/EBPbeta. J Biol Chem 2000; 275(41):32077-32088.
46. Jiang CK, Magnaldo T, Ohtsuki M et al. Epidermal growth factor and transforming growth factor alpha specifically induce the activation- and hyperproliferation-associated keratins 6 and 16. Proc Natl Acad Sci USA 1993; 90(14):6786-6790.
47. Agarwal C, Efimova T, Welter JF et al. CCAAT/enhancer-binding proteins. A role in regulation of human involucrin promoter response to phorbol ester. J Biol Chem 1999; 274(10):6190-6194.
48. Smith FJ, Jonkman MF, van Goor H et al. A mutation in human keratin K6b produces a phenocopy of the K17 disorder pachyonychia congenita type 2. Hum Mol Genet 1998; 7(7):1143-1148.
49. Chapalain V, Winter H, Langbein L et al. Is the loose anagen hair syndrome a keratin disorder? A clinical and molecular study. Arch Dermatol 2002; 138(4):501-506.
50. Wong P, Colucci-Guyon E, Takahashi K et al. Introducing a null mutation in the mouse K6alpha and K6beta genes reveals their essential structural role in the oral mucosa. J Cell Biol 2000; 150(4):921-928.
51. Mazzalupo S, Wong P, Martin P et al. Role for keratins 6 and 17 during wound closure in embryonic mouse skin. Dev Dyn 2003; 226(2):356-365.
52. Wong P, Coulombe PA. Loss of keratin 6 (K6) proteins reveals a function for intermediate filaments during wound repair. J Cell Biol 2003; 163(2):327-337.
53. Blessing M, Jorcano JL, Franke WW. Enhancer elements directing cell-type-specific expression of cytokeratin genes and changes of the epithelial cytoskeleton by transfections of hybrid cytokeratin genes. EMBO J 1989; 8(1):117-126.

CHAPTER 7

Transcriptional Regulation of Keratin Gene Expression

Miroslav Blumenberg*

Abstract

Keratin synthesis is regulated at the level of transcription. Each keratin gene appears to be regulated by a characteristic constellation of transcription factors and DNA binding sites. Often these occur in clusters and complexes, providing a mechanism for fine-tuning the expression levels. Most commonly, the important regulatory sites are found in the promoter regions, infrequently coding and downstream sequences also play a role. Transcription factors Sp1, AP1 and AP2 are important components in regulation of many keratin genes, and the nuclear receptors for retinoic acid and thyroid hormone also regulate majority of keratins. In addition, the expression of most keratin genes can be modulated by extracellular signals, such as growth factors. Universal or general regulators for all keratin genes have not been found; apparently each keratin protein has its own, characteristic circuits and machinery for regulation of expression.

Introduction

Since the first cloning of a keratin gene, just over 20 years ago,[1,2] we have come a long way toward understanding their structure, function and regulation. The control of keratin expression occurs primarily, perhaps exclusively, at the transcriptional level. Therefore, several groups have worked very hard on elucidating the transcriptional regulation of keratin gene expression. Keratins K5, K6, K16 and K18 have been particularly well explored, others, e.g., K7 and hair keratins, await their turn. A common thread that connects the transcriptional regulation of keratin genes is that their promoters contain complex sites that simultaneously bind a multitude of transcription factors. Many transcriptional factor binding sites in keratin gene promoters have been mapped, and their functional significance ascertained. However, the overall picture of the mechanisms and circuits that regulate keratin gene expression is not known. At present, we see many trees, but not yet the forest. Still, many details are known, which combined with systematic genomic studies, promises that a comprehensive picture will appear before long.

Transcriptional Regulation of Keratin K5 Expression

A common feature of all epithelial cells is the presence of keratin proteins that assemble into an intermediate filament cytoskeletal network. Whereas other cell types often use a specific master transcription factor to coordinate cell type-specific transcription, analysis of transcriptional regulation of keratin genes suggests that specific groupings of widely expressed transcription factors, acting on clusters of recognition elements in the promoter regions,

*Miroslav Blumenberg—Departments of Dermatology and Biochemistry and The Cancer Institute, NYU School of Medicine, 550 First Avenue, New York, New York, U.S.A.
Email: blumem01@med.nyu.edu

Intermediate Filaments, edited by Jesus Paramio. ©2006 Landes Bioscience and Springer Science+Business Media.

confer epithelia-specific transcription. Keratins K5 and K14 form the cytoskeletal intermediate filament network in mitotically competent basal cells in all stratified epithelia, and therefore have received extensive attention focused on the mechanisms of their regulation.

To initiate analysis of the protein factors that interact with the human K5 keratin gene upstream region, we have used gel-retardation and DNA-mediated cell-transfection assays. A cluster of three sites that binds five transcription factors was found in the promoter of the human K5 keratin gene.[3,4] Within this cluster, an unusual Sp1 site binds the Sp1 transcription factor and two additional proteins. Flanking the Sp1 site are an AP2 site and another sequence that binds a transcription factor. Similar clusters of recognition sites for the same transcription factors have been also identified in other keratin genes. DNA-protein interactions at two of the sites apparently increase transcription levels, at one decrease it, while the importance of the remaining two sites is, at present, unknown. In addition, the location of the retinoic acid and thyroid hormone nuclear receptor action site has been determined, and it involves a cluster of five sites similar to the consensus recognition elements. The complex constellation of protein binding sites upstream from the K5 gene probably reflects the complex regulatory circuits that govern the expression of the K5 keratin in mammalian tissues.[5] Such clusters may play a role in epithelia-specific expression of keratins.

In transgenic mice, as few as 90 bp of the human K5 promoter still directed expression to stratified epithelia.[6] However, the truncated K5 promoter expression was not limited to the basal layer. A 6Kb segment of the 5' upstream K5 gene directed proper basal cell-specific expression in all stratified epithelia. An open chromatin region containing a DNase I-hypersensitive site within the 6 kb of 5' upstream regulatory sequence acted independently to drive abundant and keratinocyte-specific reporter gene activity in culture and in transgenic mice. A 125-bp segment of this element is an independent strong enhancer element with keratinocyte-specific activity in vivo. Its activity is restricted to a subset of cells located within the sebaceous gland. The adjacent segment can suppress the sebocyte-specific expression and induce expression in the inner root sheath. Thus, the K5 gene expression is determined by multiple regulatory modules, which may contain AP-2 and/or Sp1/Sp3 binding sites as well as additional sites that determine cell type specificity.[7]

Similar regulation mechanism regulate K5 expression in other organism; for example, the 5' upstream region located between the cap-site and nucleotide -605 of the bovine K5 gene was found to contain the cis-regulatory DNA elements involved in the cell-type-specific expression. These elements enhanced the specific expression in the epithelial cells that express the endogenous gene, but not in cells that do not.[8] Analogous epithelium-specific expression was also observed in murine keratinocytes, suggesting that similar regulatory mechanisms have been conserved through evolution. The 5.2 kilobases preceding the gene contain the regulatory sites responsible for the cell type-specific expression of bovine K5. A strong enhancer is located between positions -762 and −1009.[9] The only regulatory element found in this enhancer by electrophoretic mobility shift, competition, and footprinting experiments is a consensus AP-1 site. Mutation of this site abolishes the activity of the enhancer and reduces to 25% the activity of the 5.2-kilobase upstream promoter region. Surprisingly, although the AP-1 sequence presents indistinguishable footprints in all cell types tested, the enhancer is active only in some of them. Furthermore, an oligonucleotide containing the AP-1 region is active in epithelial cells lines but not in fibroblasts, suggesting that this region could constitute an epithelium-specific AP-1 element. Thus the regulation of K5 keratin gene by AP-1 must be complex and different from other suprabasal, AP-1-regulated cellular and viral genes.

The tissue specificity of the bovine K5 expression was used ingeniously to produce strict conditional expression of genes in the mouse epidermis. Transgenic mouse lines were generated in which the tetracycline-regulated transcriptional transactivators, tTA and rTA, are linked to the bovine keratin 5 promoter.[10] When crossed with the tetOlacZ indicator line, the K5/tTA line induced beta-galactosidase enzyme activity in the epidermis at a level 500-fold higher than controls, and oral and topical administration of doxycycline caused a dose- and time-dependent suppression of beta-galactosidase mRNA levels and enzyme activity. Histochemical analyses localized the beta-galactosidase expression to K5 positive tissues, i.e., the basal layer of the epidermis and the outer root sheath of the hair follicle. The K5-dependent tetracycline regulatory system produces effective conditional gene expression in the mouse epidermis, and has been used to suppress and activate foreign genes specifically in the basal layer of the epidermis.

Regulation of K14

A construct containing 300 base pair segment corresponding to the promoter region of a human K14 gene was introduced into various mammalian cell lines and primary cultures.[11] The 300 base pair segment was active in all epithelial cells, including transformed simple epithelial cells, cell lines derived from stratified epithelia as well as primary cultures of epithelial cells, but it was inactive in all nonepithelial cells tested including fibroblasts and melanocytes. Using a series of deletions, an essential function was localized within a 40 bp sequence, thus identifying the keratin gene promoter that is sufficient to confer epithelial-specific expression.[11]

A much larger construct containing approximately 2.500 kb of the human K14 keratin gene generated tissue-specific and differentiation-specific expression in transgenic mice.[12] Four DNase I-hypersensitive sites are present in the 5' regulatory sequences of the K14 gene in cells where the gene is actively expressed.[13] Two of these sites are conserved in position and sequence within the human and mouse K14 genes. A novel 700-bp regulatory domain encompassing these sites is sufficient to confer epidermis-specific activity to a heterologous promoter. A 125-bp DNA fragment encompassing DNase I-hypersensitive site-II harbors the majority of the activity in vitro, and AP-1, ets, and AP-2 proteins orchestrate the keratinocyte-preferred expression.

The suprabasal suppression of the K5 and K14 gene expression may depend on the POU homeodomain factors Skn-1a and Tst-1, because K14 mRNA expression persists in suprabasal cells in Skn-1/Tst-1 double knockout mice. Both Skn-1a and Tst-1 repress the K14 promoter, with the POU domain being sufficient for repression.[14] DNA-binding defective mutants of Skn-1a and Tst-1 are as effective at mediating repression as the wild-type proteins. A 100-base pair sequence, lacking POU-binding sites, adjacent to the transcription start site of the K14 gene is sufficient and required for repression by Skn-1a, suggesting that protein-protein interactions rather than direct DNA binding effect the repression. (CBP)/p300 coactivators activate K14 gene expression and interact directly with the POU domain of Skn-1a, suggesting that POU domain factors repress K14 gene expression by interfering with CBP/p300.

When we compared the functions of epidermal keratin genes, K5, K6, K10 and K14 by transfection into nonepithelial and transformed epithelial cell lines, as well as in primary cultures of cells derived from simple and stratified epithelia, we found that the four promoters were functional only in epithelial cells.[15] While the promoter for the K14 gene was active in all epithelial cells tested, its basic-type partner, K5, and the promoter for the hyper-proliferation-associated K6 were active only in primary cultures of stratified epithelia. The promoter for the epidermal differentiation-specific K10 keratin gene was active at a low level in primary cultures of stratified epithelial cells on nonepidermal origin. Thus, the K14 gene promoter is functional in all epithelial cells, but the upstream regions of the K5 and K6 keratin genes restrict their expression to stratified epithelia, whereas the epidermal determinants of the K10 gene are not in the proximal upstream sequences.

We studied the effects of transcription factors of the AP-1 and NF-κB families on the expression of those four keratin genes using gene transfection.[16] AP-1 and NFκB are activated by many extracellular signals, including those in hyperproliferative and inflammatory processes. K5 and K14 promoters, which are coexpressed in vivo, are regulated in parallel: both were activated by the c-Fos and c-Jun components of AP-1, but not by Fra1. On the other hand, the NFκB proteins, especially p65, suppressed these two promoters. The K17 promoter was specifically activated by c-Jun, whereas the other transcription factors tested had no significant effect. In contrast, the K6 promoter was very strongly activated by all AP-1 proteins, especially by the c-Fos + c-Jun and Fra1 + c-Jun combinations. It was also strongly activated by the NFκB p65 protein. AP-1 and NFκB synergistically activated the K6 promoter, although the AP-1 and the NF-κB responsive sites could be separated physically. These results suggest that the interplay of AP-1 and NFκB proteins regulates epidermal gene expression and that the activation of these transcription factors by extracellular signaling molecules brings about the differential expression of keratin genes in epidermal differentiation, cutaneous diseases, and wound healing.

The transcription of K14 and some other epidermal marker genes is regulated by AP1 interactions at their promoters. c-Jun and JunD activate and JunB downregulates the transcription of both basal and suprabasal genes.[17] The effect of c-Jun is exerted through interactions with c-Fos at the AP1 motifs in the target promoters, whereas both JunB and JunD act independently of the binding at the AP1 sites. The differentiation specificity of the AP1 regulation seems determined by interactions involving other transcriptional regulators and transcription factor AP2 plays a role in K14 keratin gene expression. Functional AP2 binding sites were found upstream from several epidermal genes, suggesting that AP2 may be generally involved in epidermal gene regulation.[18] The role of AP2 was examined by in vitro gel shift analysis, AP2 binding site mutagenesis, and stable and transient transfection experiments.[19] Nonepithelial cells, such as fibroblasts and melanocytes, neither express keratin nor become phenotypically epithelial when transfected with an AP2-expressing vector. However, cotransfection of an AP2-expressing vector increases the level of transcription from keratin gene promoters. Thus, the role of AP2 in keratin gene expression seems to be quantitative rather than qualitative.

Regulation of K15

Basal layers of stratified squamous epithelia express keratins K15 in addition to K5 and K14, although at lower levels. Nonkeratinizing stratified epithelia, e.g., esophagus, produce more keratin K15 than epidermis. We cloned the promoter of the K15 gene and examined its regulation. Using cotransfection, gel mobility shift assays and DNAse I footprinting, we have identified the regulators of the K15 promoter activity and their binding sites.[20] We focused on those that can be manipulated with extracellular agents, transcription factors C/EBP, AP-1, NF-κB, nuclear receptors for thyroid hormone, retinoic acid and glucocorticoids, as well as the cytokine IFNγ. We found that C/EBPβ and AP1 induced, while retinoic acid and glucocorticoid receptors and NFκB suppressed the K15 promoter, along with other keratin gene promoters. However, the thyroid hormone and IFNγ uniquely and potently induced the K15 promoter.[20]

Regulation of K1 and K10

The process of epidermal differentiation is profoundly influenced by the level of intracellular calcium within keratinocytes, which, in turn, regulates the expression of the differentiation-specific keratin genes K1 and K10. A 249-bp region in the 3'-flanking region of the human K1 gene, located 7.9 kb downstream from the promoter, can activate a minimal promoter construct in transfected keratinocytes.[21] Importantly, this activity was enhanced by increased levels of calcium. The 249-bp fragment demonstrated a marked specificity for epidermal keratinocytes and was not active in other cell types. Moreover, this fragment could activate CAT expression in a construct driven by the K1 promoter, which alone had no intrinsic CAT activity. An AP-1 site is implicated in mediating the calcium response. These data identified and characterized a calcium-responsive regulatory element of the K1 gene.[21]

Forced expression of C/EBPβ in BALB/MK2 keratinocytes inhibited growth, induced morphological changes consistent with a more differentiated phenotype, and induced expression of two early markers of differentiation, K1 and K10 but had a minimal effect on the expression of late-stage markers, loricrin and involucrin.[22] Conversely, C/EBPβ-deficient mice revealed decreased expression of K1 and K10 but not of involucrin and loricrin. C/EBPβ-deficient primary keratinocytes were partially resistant to calcium-induced regulation. Thus, C/EBPβ modulates the early events of keratinocyte differentiation including K1 and K10 expression.

Similarly, overexpression of Whn (Hfh11, Foxn1), a winged-helix/forkhead transcription factor that, when mutated causes the nude phenotype, stimulates the expression of K1, while suppressing later markers, profilaggrin, loricrin, and involucrin.[23] This suggests a role for Whn, the nude gene, in the earliest stages of epithelial terminal differentiation, the stages when the expression of keratins K1 and K10 commences.

Differentiation-associated transcription factors C/EBPα, C/EBPβ, and AP-2 regulate K10 gene expression.[24] In cultured cells, C/EBPα and C/EBPβ can activate the K10 promoter using three binding sites. The selection of C/EBPα vs. C/EBPβ for K10 regulation is determined through a third transcription factor, AP-2 (see below). Unique gradients of expression exist for each transcription factor, i.e., C/EBPβ and AP-2 are most abundant in the lower epidermis, C/EBPα in the upper layers. In response to differentiation signals, loss of AP-2 expression leads to induction of C/EBPα and activation of the K10 promoter.

Regulation of K4 and K13

The esophageal stratified epithelium, like the epidermis, comprises an actively proliferating basal layer that, unlike the epidermis, undergoes a differentiation program not involving keratinization. Given its localization, K4 is a marker of the early differentiated suprabasal compartment of nonkeratinizing stratified epithelia. The transcriptional regulatory signals that orchestrate the switch from proliferation to differentiation also regulate the human K4 promoter.[25] A critical cis-regulatory element contains an inverted CACACCT motif that binds esophageal-specific zinc-dependent transcription factors. Importantly, the interaction between Sp1 and cell cycle regulatory proteins is important in regulating K4 expression.[26] Sp1 activation of the K4 promoter was reduced in cyclin D1-overexpressing cells, which is possibly mediated through direct interaction between Sp1 and cyclin D1; the reduction is opposed by a complex between Sp1 and pRB.

AP-2 also transactivates K4 transcription. The promoter region of K4 contains a functional AP-2 binding site in the vicinity of the transcriptional start.[27] Various constructs, which did or did not contain the K4 promoter AP-2 site, were ballistically transfected into differentiating HaCaT keratinocytes. The results revealed that the AP-2 site is functional, although additional regulatory elements were found to be necessary for the full transcription of K4. These include the Kruppel-like transcriptional factors KLF6 and KLF4, which physically interact and coactivate the K4 promoter.[28] KLF6 is a widely expressed member of the Kruppel-like family found in the esophagus. Using transient transfection, KLF6 was found to transactivate the human gene K4 promoter; cotransfection of KLF6 and KLF4, another member of the Kruppel-like factors and expressed in the esophageal squamous epithelium, leads to additive activation. The promoter contains a CACCC-like motif previously shown to bind KLF4. In a transient transfection system, KLF4 increased the activity of K4 promoter >25-fold, which depended on the CACCC-like element.[29]

The ets transcription factors contribute to diverse cellular functions and include a novel epithelial-specific member of the ets family ELF3 (a.k.a. ESE-1, ERT, jen, ESX). Interestingly, ELF3 suppressed the basal expression of K4 promoter in both esophageal and cervical epithelial cancer cell lines, while simultaneously activating the SPRR2A promoter, linked to late-differentiation.[30] ELF3 may have dual functions in the transcriptional regulation of squamous epithelial differentiation, including transcriptional suppression of the K4 promoter.

K13, together with K4, its basic partner, is expressed in the suprabasal layers of noncornified stratified epithelia. Sequence analysis has revealed that two transcription-start sites were utilized, the major being at 61 and the minor at 63 nucleotides upstream of ATG.[31] The promoter contains a TATA box and several other putative transcription factor binding sites. K13 is aberrantly expressed in murine epidermal tumors and constitutes an early marker of malignant progression.[32] In vitro, expression of K13 in transformed epidermal cell lines can be induced either by Ca^{2+} or, indirectly, by retinoic acid.

Regulation of K3 and K12

Corneal epithelial cells initially express K5 and K14 keratins, characteristic of basal keratinocytes, and then undergo biochemical differentiation, as evidenced by the subsequent expression of K3 and K12 keratin markers of corneal epithelial differentiation. Using rabbit corneal epithelial cells, the promoter region of the rabbit K3 promoter was analyzed in transfection experiments. Serial deletion experiments narrowed this keratinocyte-specific promoter to within -300 bp upstream of the transcription initiation site. Its activity was not regulated by the coding or 3'-noncoding sequences. This 300 bp sequence can function in vitro as a keratinocyte-specific promoter and contains two clusters of partially overlapping motifs, one with an NFκB consensus sequence and another with a GC box.[33] The combinatorial effects of these multiple motifs and their cognate binding proteins may play an important role in regulating the expression of this tissue-restricted and differentiation-dependent keratin gene. Electrophoretic mobility shift assays established that corneal keratinocyte nuclear proteins bind in vitro to the two sites. Immunosupershift and UV cross-linking show that NFκB, consisting of the p65 and p50 subunits, bind to the sequence GGGGCTTTCC, -262 to -253 bp. The second site contain unusual overlapping Sp1 and AP-2 GC-rich motif, CCGCCCCCTG, at -203 to -194 bp. This site bound an Sp1-related keratinocyte nuclear protein. Mutagenesis of the NFκB site, GC motif, and both abolished 20, 50, and 75% of the promoter activity, respectively, in transfected keratinocytes. These results indicate that NFκB is present in significant quantities in keratinocyte nuclei and that the tissue restriction of the NFκB- and Sp1-related proteins, in combination with other factors, may contribute to the keratinocyte specificity of rabbit K3 promoter. Furthermore, Sp1 activates, while AP-2 represses the K3 promoter.[34] Although corneal basal cells express approximately equal amounts of Sp1 and AP-2 DNA-binding activities, the differentiated cells drastically down-regulate their AP-2 activity, which results in a six to sevenfold increase of the Sp1/AP-2 ratio. This change coincides with the activation of the differentiation-related K3 gene and suppression of the K14 keratin gene. In addition, polyamines, which are present in a high concentration in proliferating basal keratinocytes, inhibit the binding of Sp1 to its binding motif, but do not inhibit the binding of AP-2. These results suggest that the low Sp1/AP-2 ratio and the polyamine-mediated inhibition of Sp1 may account for the suppression of the K3 gene expression in the corneal basal cells, while the elevated Sp1/AP-2 ratio may be activating the K3 gene in the differentiated corneal epithelial cells. The switch of the Sp1/AP-2 ratio during corneal epithelial differentiation apparently plays a role in the reciprocal expression of the K3 and K14 genes during corneal differentiation.

K12 is the acidic type pair of K3 in differentiated corneal epithelial cells. The 2.5-kb DNA fragment 5'-flanking K12 contains corneal epithelial cell-specific regulatory cis-DNA elements. Pax-6 is a positive transcription factor essential for K12 expression. Three 5' truncated fragments of the keratin K12 promoter (1.03, 0.71 and 0.25 Kb) showed higher functional and tissue-specific promoter activity in a human corneal epithelial cell line than other cell lines.[35] The 250 bp K12 promoter fragment was active in cultured rabbit corneal epithelial cells, suggesting that the tissue-specific expression in corneal epithelial cells extends across species lines. The paired box homeotic gene 6 (PAX-6), which plays a role in controlling eye development, stimulates the activity of keratin K12 promoter.[36] The cis-regulatory elements located 600 bp upstream of the murine K12 gene were analyzed in rabbit corneas using particle-mediated gene transfer "Gene Gun" technology, while DNA foot-printing and electrophoresis mobility shift assay were performed to identify the cis-regulatory elements using bovine corneal epithelial cell nuclear extracts. The sequences between -181 to -111 and -256 to -193 bound to nuclear proteins; these two regions were potential binding sites for many transcription factors, such as AP1, c/EBP, and KLF6.[37]

Regulation of Hair Keratins

During hair growth, cortical cells differentiate and synthesize large amounts of hair keratin proteins. Hair keratin gene regulation has been studied in mouse and humans, as well as in sheep, where wool production has major economic consequences. The organogenesis of hair follicles is a very complex process, regulated both positively and negatively by several members of the Delta/Notch signaling system.[38,39]

HOXC13 when mutated or overexpressed in mice produces a fragile hair phenotype. HOXC13, but not a homeobox-deleted HOXC13 mutant, strongly activated the promoters of coexpressed human hair keratin genes.[40] The hair keratin promoters contained numerous putative Hox binding core motifs TAAT, TTAT, and TTAC. Electrophoretic mobility shift assays identified the core motifs concentrated in the proximal promoter regions and allowed the deduction of an HOXC13 consensus binding sequence TT(A/T)ATNPuPu. Thus, there seems to be a direct involvement of HOXC13 in the control of hair keratin expression during early trichocyte differentiation.

Minimal promoter of the wool keratin gene K2.10 spanning nucleotides -350 to +53 was sufficient to direct expression of the lacZ gene to the hair follicle cortex of transgenic mice.[41] Mutation introduced into the binding site for lymphoid enhancer factor 1 (LEF-1) decreased promoter activity, without affecting specificity. The constituent proteins of the LEF/TCF transcription complexes change during hair follicle differentiation; the LEF/TCF complexes seem to regulate directly the expression of hair keratin genes.[42] DNase I footprinting analyses and electrophoretic mobility shift assays identified LEF-1, Sp1, AP2-like and NF1-like proteins bound to the promoter. The LEF-1 binding site is an enhancer element of the K2.10 promoter in the hair follicle cortex and additional factors may regulate the tissue- and differentiation-specificity of the promoter.

Regulation of Keratin Gene Expression by Hormones and Vitamins

Nuclear receptors for retinoic acid and thyroid hormone regulate transcription of keratin genes.[43] By cotransfecting the vectors expressing nuclear receptors for retinoic acid, RAR, and thyroid hormone, T3R, along with the constructs that contain K5, K6, K10 and K14 gene promoters into epithelial cells, we have demonstrated that the receptors can suppress the promoters of keratin genes. The suppression is ligand dependent. The regulation of keratin gene expression by RAR and T3R occurs through direct binding of these receptors to the receptor response elements of the keratin gene promoters.[44] The DNA binding domains of the receptors are essential for regulation, but the NH2-terminal "A/B" domains are not required. These findings indicate that the mechanism of regulation of keratin genes by RAR and T3R differs significantly from the mechanisms described for other genes modulated by these receptors. The RAR and T3R binding sites in the K5 promoter are adjacent to each other whereas in the K15 and K17 promoters they overlap.[20]

Similar and confirming results were obtained with squamous carcinoma cells of head and neck, where receptor-specific ligands implicated RARβ in inhibition of keratinization and suppression of K1 keratin expression.[45] Curiously, although vitamin D3 acts through a nuclear receptor, VDR, a member of the RAR/T3R family, and regulates keratinocyte differentiation, VDR does not directly interact with the keratin genes.[46] The keratin promoters shown not to be regulated by vitamin D3 include K3, K5, K10, K14, and K16.

We identified the RAR- and T3R-responsive site in the K14 gene using site-specific mutagenesis and found that the site consists of a cluster of consensus palindrome half-sites in various orientations, relatively close to the TATA box.[47] Similar clusters of half-sites that share structural organization with the K14 regulatory site were found in the K5, K6, K10, K15, K16, and K17 keratin gene promoters and are responsible for retinoic acid-mediated transcription regulation of keratin synthesis in the epidermis.[48] This means that the clustered structure of the RAR/T3R responsive elements is a common characteristic for keratin genes. Furthermore, in the absence of the ligand, T3R activates keratin gene expression, and the heterodimerization with the retinoid X receptor is not essential for activation by the unliganded T3R. Unlike other nuclear receptor binding sites, the response elements in keratin genes bind RAR, T3R and the glucocorticoid receptor, GR, as monomers or homodimers.[49] Interestingly, addition of ligand to the receptor changes the binding pattern of the T3R from homodimer to monomer, reflecting the change in regulation from induction to inhibition. Such specific DNA-receptor interactions are crucial for the repression signal of transcription. Thus, they not only provide a docking platform for the receptors, but also directing the receptors to bind in a particular configuration and coordinate the interactions among different receptors. Furthermore, the response elements allow simultaneous binding of multiple receptors, thus providing fine-tuning of transcriptional regulation.

Glucocorticoids are important regulators of epidermal growth, differentiation, and homeostasis, and are used extensively in the treatment of skin diseases. GC action is mediated *via* GR. Transgenic mice overexpressing GR in epidermis have altered skin development and impaired proliferative and inflammatory responses.[50] The developmental and proliferative phenotype includes lesions that range from epidermal hypoplasia to a complete absence of epidermis. Additional abnormalities resemble the clinical findings in patients with ectodermal dysplasia, including *aplasia cutis congenitalis*. The anti-inflammatory role of glucocorticoids is also evident in the transgenic animals, and is partly due to interference with other transcription factors, such as NFκB.[50] We have described a novel mechanism of keratin gene regulation in skin through glucocorticoid receptor monomers.[51] Glucocorticoids repress the expression of a subset of keratin genes, the basal-cell-specific keratins K5 and K14 and disease-associated keratins K6, K16, and K17, but not the differentiation-specific keratins K3 and K10 or the simple epithelium-specific keratins K8, K18, and K19. The regulated keratins are all associated with diseases, and in this way glucocorticoids differ from retinoic acid and thyroid hormone, which regulate all keratin genes tested. Detailed footprinting analysis revealed that the GR binds as four monomers to adjacent sites in keratin gene promoters.[51] Using cotransfection and antisense technology we have found that coregulators SRC-1 and GRIP-1 are not involved in the suppression of keratin genes, while histone acetyltransferase and CBP are, which is another unusual and keratin gene-specific aspect of regulation by GR. In addition, GR blocks the induction of keratin gene expression by AP1 independently from direct binding of AP1 proteins to their responsive elements. Therefore GR suppresses keratin gene expression through two independent mechanisms: directly, through interactions of four GR monomers with responsive elements in keratin genes, as well as indirectly, by blocking the AP1-mediated induction of keratin gene expression.

A large variety of signals modulates epidermal keratin gene expression. Hormones and vitamins, which act via nuclear receptors, affect the differentiation process, whereas growth factors and cytokines, which act via cell surface receptors (see below), affect keratinocyte activation and related events. We examined the interaction between the nuclear receptor and cell surface receptor pathways in regulating the expression of keratin genes.[52] Expecting to find dominance of one of the pathways, we were surprised to find that the two pathways are codominant. While EGF induces and retinoic acid suppresses expression of K6 and K16 keratin genes, when both EGF and retinoic acid are present simultaneously, the level of expression is intermediate. Similar codominant effects were found on other keratin genes with IFNγ, TGFβ, and thyroid hormone signaling. A judicious combination of hormones, vitamins, growth factors, and cytokines may be used to target specific expression of appropriate genes in the treatment of human epidermal diseases.[20]

Regulation of Keratin Gene Expression by Growth Factors and Cytokines

Keratin K17, the myoepithelial keratin, while not present in healthy skin, is expressed under various pathological conditions. Psoriasis is associated with production of IFNγ. The primary molecular effect of IFNγ is activation of specific transcription factors, such as STAT1, which regulate gene expression in target cells. Induction of cutaneous delayed-type hypersensitivity reactions resulted in activation and nuclear translocation of STAT1 in keratinocytes in vivo and subsequent induction of transcription of keratin K17. Within the promoter of the K17 keratin gene, we have identified the site that confers the responsiveness to IFNγ and that binds the transcription factor STAT1, and thus characterized at the molecular level the signaling pathway produced by the infiltration of lymphocytes in skin and resulting in the specific induction of K17 keratin gene expression in keratinocytes.[53] The induction of K17 is specific for the inflammatory reactions associated with high levels of IFNγ and activation of STAT1, such as psoriasis and dermatitis caused by delayed type hypersensitivity, but not in samples of atopic dermatitis, which is not.[54] Two cytokines, interleukin-6 and leukemia inhibitory factor, which can induce phosphorylation of STAT1, can also induce K17 expression, whereas interleukin-3, interleukin-4, interleukin-10, and granulocyte macrophage colony stimulating factor have no effect on K17 expression. Therefore, in inflammatory skin diseases, lymphocytes, through the cytokines they produce, differently regulate not only each other, but also keratin gene expression in epidermis one of their target tissues.

In the promoter region of the K17 gene we identified eight protein binding sites.[55] Five of them bind the known transcription factors NF1, AP2, and Sp1 and three bind still unidentified proteins. Using site-directed mutagenesis, we have demonstrated the importance of the protein binding sites for the promoter function involved in both constitutive and IFNγ-induced expression of the K17 keratin gene. Interestingly, UVA irradiation (320-400 nm), but not UVB (290-320 nm), induced an increase in K17, showing a differential gene regulation between these two ultraviolet ranges.[56] Importantly, UVB, a common damaging agent in the epidermis, increased the transcription of K19 gene and to a lesser extent the K6, K5, and K14 genes.[56]

In contrast, TGFβ causes transcriptional induction of K5 and K14 keratin genes.[57] No other keratin gene promoters were induced. TGFβ is an important regulator of epidermal keratinocyte function because it suppresses cell proliferation. The effect of TGFβ is concentration-dependent, can be demonstrated in HeLa cells, does not depend on keratinocyte growth conditions and can be elicited by both TGFβ1 and TGFβ2. These results suggest that TGFβ promotes the basal cell phenotype in stratified epithelia such as the epidermis, and that the effects of TGFβ are not anti-proliferative, but merely anti-hyperproliferative.

We will not go into details of the analysis of the activation-specific keratin genes K6 and K16, which have been studied by many investigators,[58-64] and which are described elsewhere in this issue (see chapter by Navarro et al). We will briefly focus on the effects of proliferative and proinflammatory cytokines on K6 and K16 expression. For example, Interleukin-1 (IL-1) and tumor necrosis factor-α (TNFα) induce K6b keratin synthesis through a transcriptional complex containing NFκB and C/EBPβ,[65,66] while the epidermal growth factor (EGF) and transforming growth factor-α (TGFα) specifically induce K6 and K16, apparently through AP1 transcription elements.[67]

Regulation of Small Keratins, K7, K8, K18 and K19

So far, very little is known about the transcriptional regulation of K7, although a systematic exploration of its expression in several organisms has begun and, hopefully, will soon yield much more data.[68] K7 is expressed observed in lung, bladder, mesothelium, hair follicle, filiform papillae of the tongue and in a range of "hard" epithelial tissues.

The human keratins K8 and K18 genes are expressed in diverse simple epithelial tissues and in various carcinomas. Relatively little is known about the regulation of the basic type simple epithelial keratin K8, originally known as EndoA.[69,70] Its expression is regulated by p53 through a binding site in the 5' untranslated region of the gene.[71] In addition, an enhancer at the 3'-end of the gene contains seven Ets binding sites that bind ETS1, ETS2 and ERGB/FLI-1 transcription factors and regulate transcription in combination with the K8 promoter.[71,72]

In contrast, the regulation of expression of the K18 gene has received much attention because of several very exciting features unique to this gene. In transgenic mice, a 10-kilobase DNA segment of the human K18 gene contains all the necessary information for proper tissue-specific expression. Furthermore, this expression is copy number-dependent and integration site-independent. The 10 Kb sequence contains several interdependent regulatory sites, including a promoter, an intronic enhancer, a regulatory site in the exon, and the locus controlling regions, LCRs, bracketing the gene.

The promoter of the K18 gene is regulated in a cell-type specific manner. The "minimal promoter" contains the TATA box and an initiation site, and the TATA box is the only essential element of the minimal promoter.[73] The K18 expression is different in tumorigenic and nontumorigenic cell lines. The differential expression depends on an Sp1 site close upstream from the TATA box. Three different proteins bind to the Sp1 site, one of them is Sp1 itself. Importantly, the acetylation state of the proteins bound to DNA in the K18 promoter greatly affect the promoter activity: histone deacetylase inhibitors stimulate the activity of the K18 promoter.[74] CBP/p300 coactivator protein seems involved in the regulation of K18 promoter activity, at least in nontumorigenic cell lines.[75]

The promoter activity is increased by an enhancer element in intron 1 (see below). The intronic enhancer can also stimulate transcription from a cryptic promoter, apparent in the absence of the minimal promoter. Furthermore, this cryptic promoter sequence is AP-1-dependent.[76] Apparently, the minimal promoter, the cryptic promoter/enhancer, and the intron-1 enhancer interact in a very complex way, involving AP1, Sp1, CBP/p300 and protein acetylation to regulate the levels of activity of the K18 promoter.

This complexity, however, can be exploited to deliver transgenes to the airway epithelia, which is potentially very useful for development of gene therapy approaches.[77] A helper-dependent adenoviral vectors utilizing the K18 sequence as a tissue-specific promoter, directed expression of linked genes in airway epithelial and submucosal cells, apparently circumventing the side effects such as acute toxicity and inflammation, and could be important in the development of gene therapy for cystic fibrosis

The induction of K18 in embryonal carcinoma (EC) and embryonic stem (ES) cells can be triggered in culture by exposure to retinoic acid. The indiction depends, in part, on the complex enhancer element located within the first intron of K18.[78] ETS-2 and AP-1 transcription factor c-Jun, and JunB also mediate the induction of K18. These transcription factors act by opposing three silencer elements also located within the first intron of the K18 gene. Therefore, the induction of K18 is due to a combination of relief from negative regulation and direct positive activation by the ETS and AP-1 transcription factors. The methylation of the ETS binding site causes a repression of the K18 gene expression, because ETS proteins do not bind to methylated sites in DNA. Such methylation adds another layer of complexity to the regulation of K18 keratin gene expression.[79]

DNase-hypersensitive sites often correlate with regulatory regions of genes. Four such sites are found in the promoter region and the first intron of K18. Two DNase-hypersensitive sites were found in an Alu repetitive sequence immediately upstream of the promoter.[80] The final hypersensitive site was mapped to exon 6 of K18. This novel regulatory element in exon 6 modulates K18 gene expression in transgenic mice. While the exon 6 site can bind c-Jun and c-Fos, it can stimulate transcription independent of AP-1 proteins.[81] Therefore, the protein-coding sequences of the K18 gene have a regulatory function as well.

Arguably the most unusual and least understood DNA regulatory sites are the locus control regions, LCRs.[82] These sequences, first described bracketing the hemoglobin genes, confer integration site-independent transcription in transgenic mice. LCR serves to insulate the region from transcriptional influences of surrounding DNA. Neither read-through, nor enhancers nor silencers have any effect in the segments protected by LCRs. This means that the level of expression of a transgene depends only on the copy number, and not on the site of integration.

The group of Robert Oshima identified and characterized the LCR elements of the human K18 keratin gene. Transgenic animals express the human K18 gene at levels proportional to the copy number and independently of the integration sites.[83,84] Integration site-independent expression of K18 depends on a 323 bp sequence from the 5' flanking sequence and 3.5 kb of the 3' flanking sequence. The LCR activity is orientation-dependent. Interestingly, the integration site-independent expression and copy number-dependent expression can be separated. The 323 bp sequence can confer LCR function even on heterologous transgenes.[85] This sequence comprises an Alu repetitive element and its LCR activity correlates with its RNA polymerase III promoter activity.

Keratin K19 is expressed in simple and nonkeratinizing stratified epithelia. Its expression seems associated with malignant transformation in esophageal and pancreatic cancers. This expression depends on a sequence containing binding sites for KLF4 and Sp1. Furthermore, overexpression of KLF4 and Sp1 induces K19 production.[86] An enhancer sequence containing an AP1 site has been identified in the K19 3'-flanking region.[87] K19 is specifically and directly induced by retinoic acid[88] and by estrogen.[89] The estrogen regulation depends on a complex enhancer region in the first intron, which contains several estrogen receptor binding sites and AP1 sites.

The K19 upstream sequences contain transcription regulatory elements in -2249 to -2050 bp and -732 bp to the first ATG. Six protein-binding sites, including an Sp1 site, a CCAAT box and a TATA box were detected in the segment from -732 bp to the first ATG by the DNA footprinting technique.[90] The K19 promoter has been used for targeting and identification of K19-producing cells in transgenic mice.[91] Expression was found in ductal epithelia of the pancreas, small intestinal villi, in surface colonocytes, and gastric isthmus cells. The expression of K19 correlated with and overlapped that of KLF4 protein expression.

Specific Transcription Factors Implicated in Regulation of Transcription of Keratin Genes

The transcription factors AP-2 comprise three isoforms, AP-2-α, -β and -γ, similar in structure. AP-2α is present in basal keratinocytes, but is significantly induced in proliferating diseases; AP-2β is present only in sweat glands, whereas AP-2γ is present throughout the epidermis in normal and psoriatic skin as well as in the SCC lesions. The K14 promoter binds to AP-2α and AP-2γ, whereas the K1 promoter predominantly binds to AP-2γ. In contrast, AP-2β does not bind to either keratin DNA.[92] The AP2 proteins may be general regulators of keratin gene expression,[18,19] but their precise role remains to be elucidated.

Ectopic expression of c-Myc,[93] KLF4,[94] Gli2[95] or Ras and Raf[96] in the epidermis alters cell phenotype and keratin expression. However, the effects of these transcription factors on keratin gene expression may be indirect. Particularly interesting are the roles of the Kruppel-like transcription factors.[94] KLF4, besides affecting epidermal barrier function in development, directly regulates K4 and K19 (see above).[28,86] On the other hand, K12, the corneal keratin gene is regulated by KLF6, presumably through the KLF6 binding site identified in the human K12 promoter.[97]

Table 1. Transcription factors shown to regulate keratin genes

Keratin	Sp1	AP1	RAR/T3R	GR	ets	KLF4	AP2	C/EBP	NFκB	CBP/p300	Others
											Transcription Factors
K1		Yes	Yes					Yes			Whn
K3	Yes	Yes					Yes		Yes		
K4	Yes	Yes			Yes	Yes	Yes				KLF6
K5	Yes	Yes	Yes	Yes			Yes		Yes		
K6			Yes	Yes				Yes	Yes		
K8					Yes						
K10			Yes				Yes	Yes			
K12		Yes	Yes					Yes			KLF6, PAX-6
K14		Yes	Yes	Yes	Yes		Yes		Yes		Skn-1a and Tst-1
K15		Yes	Yes	Yes			Yes	Yes	Yes	Yes	STAT1
K16			Yes	Yes				Yes	Yes		
K17	Yes		Yes	Yes			Yes				STAT1
K18	Yes	Yes	Yes		Yes					Yes	
K19	Yes		Yes			Yes					
Hair K2.10	Yes						Yes				HOX-13, LEF/CTF, NF1

Conclusions and Future Directions

The common characteristic of all keratin gene promoters analyzed so far indicates that each individual gene is regulated by numerous transcription factors that bind nearby DNA elements and assemble into multi-protein complexes. These transcription factors respond to extracellular influences, such as hormones, vitamins and growth factors. Therefore, the level of expression of each keratin gene seems to be affected by the signals from the cellular milieu.

Conspicuously absent is the "epithelia-specific master regulator", a transcription factor common to all epithelial cells and specific only for epithelial cells, necessary for expression of all keratin genes. Hope for finding such a protein derives from the axiomatic association of keratins and epithelia. Its absence means that every epithelial cell type has elaborated a set of transcription factors that, in combinations, are responsible for expression of keratins. Furthermore, we did not find "epidermal", "simple epithelial" or even "differentiation-specific" master regulators that are either permissive or necessary for expression of a subclass of keratins.

Similarly, while keratin proteins tend to be expressed in specific pairs that consist of a basic and an acidic keratin, the pair-wise coexpression regulators have not been identified. Again, we are left with the notion that each keratin gene is regulated by its own regulatory circuits and expressed independently from other keratins. The pair-wise coexpression, then, is a fortuitous result of multiple independent regulatory mechanisms.

On the other hand, the frequent encounter of a small number of transcription factors in the regulation of many keratin genes indicates that same regulatory motifs play a role in the expression of keratin genes. The recently elucidated sequences of several mammalian species will soon lead to new genomics breakthroughs. Before long, algorithms for large throughput sequence analysis of promoters will be available. Hopefully, many more discoveries and surprises are in store for keratinologists in the near future.

Acknowledgements

Our experiments have been supported by grant AR41850 from the National Institutes of Health. I want to thank my collaborators over the past years, especially Marjana Tomic-Canic, Alix Gazel, Chun-Kui Jiang, Tomohiro Banno and Mayumi Komine, among others, for joining me on different legs of this exciting and fun journey.

References

1. Steinert PM, Rice RH, Roop DR et al. Complete amino acid sequence of a mouse epidermal keratin subunit and implications for the structure of intermediate filaments. Nature 1983; 302(5911):794-800.
2. Hanukoglu I, Fuchs E. The cDNA sequence of a type II cytoskeletal keratin reveals constant and variable structural domains among keratins. Cell 1983; 33(3):915-24.
3. Ohtsuki M, Tomic-Canic M, Freedberg IM et al. Nuclear proteins involved in transcription of the human K5 keratin gene. J Invest Dermatol 1992; 99:206-15.
4. Ohtsuki M, Flanagan S, Freedberg IM et al. A cluster of five nuclear proteins regulates keratin transcription. Gene Expr 1993; 3:201-13.
5. Ohtsuki M, Tomic-Canic M, Freedberg IM et al. Regulation of epidermal keratin expression by retinoic acid and thyroid hormone. J Dermatol 1992; 19(11):774-80.
6. Byrne C, Fuchs E. Probing keratinocyte and differentiation specificity of the human K5 promoter in vitro and in transgenic mice. Mol Cell Biol 1993; 13(6):3176-90.
7. Kaufman CK, Sinha S, Bolotin D et al. Dissection of a complex enhancer element: Maintenance of keratinocyte specificity but loss of differentiation specificity. Mol Cell Biol 2002; 22(12):4293-308.
8. Blessing M, Jorcano JL, Franke WW. Enhancer elements directing cell-type-specific expression of cytokeratin genes and changes of the epithelial cytoskeleton by transfections of hybrid cytokeratin genes. EMBO J 1989; 8(1):117-26.
9. Casatorres J, Navarro JM, Blessing M et al. Analysis of the control of expression and tissue specificity of the keratin 5 gene, characteristic of basal keratinocytes. Fundamental role of an AP-1 element. J Biol Chem 1994; 269(32):20489-96.
10. Diamond I, Owolabi T, Marco M et al. Conditional gene expression in the epidermis of transgenic mice using the tetracycline-regulated transactivators TTA and RTA linked to the keratin 5 promoter. J Invest Dermatol 2000; 115(5):788-94.

11. Jiang CK, Epstein HS, Tomic M et al. Epithelial-specific keratin gene expression: Identification of A 300 base-pair controlling segment. Nucleic Acid Res 1990; 18(2):247-53.
12. Vassar R, Rosenberg M, Ross S et al. Tissue-specific and differentiation-specific expression of a human K14 keratin gene in transgenic mice. Proc Natl Acad Sci USA 1989; 86(5):1563-7.
13. Sinha S, Degenstein L, Copenhaver C et al. Defining the regulatory factors required for epidermal gene expression. Mol Cell Biol 2000; 20(7):2543-55.
14. Sugihara TM, Kudryavtseva EI, Kumar V et al. The pou domain factor skin-1a represses the keratin 14 promoter independent of DNA binding. A possible role for interactions between Skn-1a and CREB-binding protein/P300. J Biol Chem 2001; 276(35):33036-44.
15. Jiang CK, Epstein HS, Tomic M et al. Functional comparison of the upstream regulatory DNA sequences of four human epidermal keratin genes. J Invest Dermatol 1991; 96:162-67.
16. Ma S, Rao L, Freedberg IM et al. Transcriptional control of K5, K6, K14, and K17 keratin genes by AP1 and NF-kappaB family members. Gene Expr 1997; 6(6):361-70.
17. Rossi A, Jang SI, Ceci R et al. Effect of AP1 transcription factors on the regulation of transcription in normal human epidermal keratinocytes. J Invest Dermatol 1998; 110(1):34-40.
18. Leask A, Byrne C, Fuchs E. Transcription factor AP2 and its role in epidermal-specific gene expression. Proc Natl Acad Sci USA 1991; 88:7948-52.
19. Magnaldo T, Vidal RG, Ohtsuki M et al. On the role of AP2 in epithelial-specific gene expression. Gene Expr 1993; 3(3):307-15.
20. Radoja NS, Waseem O, Tomic-Canic A et al. Thyroid hormone and interferon-gamma specifically increase K15 gene transcription. Moll Cell Biol 2004; In Press.
21. Rothnagel JA, Greenhalgh DA, Gagne TA et al. Identification of a Calcium-inducible, epidermal-specific regulatory element in the 3'-flanking region of the human keratin 1 gene. J Invest Dermatol 1993; 101:506-13.
22. Zhu S, Oh HS, Shim M et al. C/EBPbeta modulates the early events of keratinocyte differentiation involving growth arrest and keratin 1 and keratin 10 expression. Mol Cell Biol 1999; 19(10):7181-90.
23. Baxter RM, Brissette JL. Role of the nude gene in epithelial terminal differentiation. J Invest Dermatol 2002; 118(2):303-9.
24. Maytin EV, Lin JC, Krishnamurthy R et al. Keratin 10 gene expression during differentiation of mouse epidermis requires transcription factors C/EBP and AP2. Dev Biol 1999; 216(1):164-81.
25. Opitz OG, Rustgi AK. Interaction between Sp1 and cell cycle regulatory proteins is important in transactivation of a differentiation-related gene. Cancer Res 2000; 60(11):2825-30.
26. Opitz OG, Jenkins TD, Rustgi AK. Transcriptional regulation of the differentiation-linked human K4 promoter is dependent upon esophageal-specific nuclear factors. J Biol Chem 1998; 273(37):23912-21.
27. Wanner R, Zhang J, Dorbic T et al. The promoter of the HACAT keratinocyte differentiation-related gene keratin 4 contains a functional AP2 binding site. Arch Dermatol Res 1997; 289(12):705-8.
28. Okano J, Opitz OG, Nakagawa H et al. The Kruppel-like transcriptional factors Zf9 and Gklf coactivate the human keratin 4 promoter and physically interact. Febs Lett 2000; 473(1):95-100.
29. Jenkins TD, Opitz OG, Okano J et al. Transactivation of the human keratin 4 and Epstein-barr virus Ed-L2 promoters by Gut-Enriched Kruppel-like factor. J Biol Chem 1998; 273(17):10747-54.
30. Brembeck FH, Opitz OG, Libermann TA et al. Dual function of the Epithelial specific ets transcription factor, Elf3, in modulating differentiation. Oncogene 2000; 19(15):1941-9.
31. Waseem A, Alam Y, Dogan B et al. Isolation, sequence and expression of the gene encoding human keratin 13. Gene 1998; 215(2):269-79.
32. Winter H, Fink P, Schweizer J. Retinoic acid-induced normal and tumor-associated aberrant expression of the murine keratin K13 gene does not involve a promotor sequence with striking homology to a natural retinoic acid responsive element. Carcinogenesis 1994; 15(11):2653-6.
33. Wu RL, Galvin S, Wu SK et al. A 300 Bp 5'-upstream sequence of a differentiation-dependent rabbit K3 keratin gene can serve as a keratinocyte-specific promoter. J Cell Sci 1993; 105(Pt 2):303-16.
34. Chen TT, Wu RL, Castro-Munozledo F et al. Regulation of K3 keratin gene transcription by Sp1 and AP2 in differentiating rabbit corneal epithelial cells. Mol Cell Biol 1997; 17(6):3056-64.
35. Liu JJ, Kao WW, Wilson SE. Corneal epithelium-specific mouse keratin K12 promoter. Exp Eye Res 1999; 68(3):295-301.
36. Shiraishi A, Converse RL, Liu CY et al. Identification of the cornea-specific keratin 12 promoter by in vivo particle-mediated gene transfer. Invest Ophthalmol Vis Sci 1998; 39(13):2554-61.
37. Wang IJ, Carlson EC, Liu CY et al. Cis-regulatory elements of the mouse Krt1.12 gene identification of the cornea-specific keratin 12 promoter by in vivo particle-mediated gene transfer. Mol Vis 2002; 8(13):94-101.

38. Lin MH, Leimeister C, Gessler M et al. Activation of the Notch Pathway in the hair cortex leads to aberrant differentiation of the adjacent hair-shaft layers. Development 2000; 127(11):2421-32.
39. Niemann C, Owens DM, Hulsken J et al. Expression of deltanlef1 in mouse epidermis results in differentiation of hair follicles into squamous epidermal cysts and formation of skin tumours. Development 2002; 129(1):95-109.
40. Jave-Suarez LF, Langbein L, Winter H et al. Androgen regulation of the human hair follicle: The type I hair keratin Hha7 is a direct target gene in trichocytes. J Invest Dermatol 2004; 122(3):555-64.
41. Dunn SM, Keough RA, Rogers GE et al. Regulation of hair gene expression. Exp Dermatol 1999; 8(4):341-2.
42. Dasgupta R, Fuchs E. Multiple roles for activated Lef/Tcf transcription complexes during hair follicle development and differentiation. Development 1999; 126(20):4557-68.
43. Tomic M, Jiang C-K, Epstein HS et al. Nuclear receptors for retinoic acid and thyroid hormone regulate transcription of keratin genes. Cell Regul 1990; 1:965-73.
44. Tomic-Canic M, Day D, Samuels HH et al. Novel regulation of keratin gene expression by thyroid hormone and retinoid receptors. Submitted 1995.
45. Zou CP, Hong WK, Lotan R. Expression of retinoic acid receptor beta is associated with inhibition of keratinization in human head and neck squamous carcinoma cells. Differentiation 1999; 64(2):123-32.
46. Blumenberg M, Connolly DM, Freedberg IM. Regulation of keratin gene expression: The role of the nuclear receptors for retinoic acid, thyroid hormone and vitamin D3. J Invest Dermatol 1992; 98:42s-49s.
47. Tomic-Canic M, Sunjevaric I, Freedberg IM et al. Identification of the retinoic acid and thyroid hormone receptor-responsive element in the human K14 keratin gene. J Invest Dermatol 1992; 99:842-47.
48. Radoja N, Diaz DV, Minars TJ et al. Specific organization of the negative response elements for retinoic acid and thyroid hormone receptors in keratin gene family. J Invest Dermatol 1997; 109(4):566-72.
49. Jho SH, Radoja N, Im MJ et al. Negative response elements in keratin genes mediate transcriptional repression and the cross-talk among nuclear receptors. J Biol Chem 2001; 276(49):45914-20, Epub 2001 Oct 8.
50. Perez P, Page A, Bravo A et al. Altered skin development and impaired proliferative and inflammatory responses in transgenic mice overexpressing the Glucocorticoid receptor. FASEB J 2001; 15(11):2030-2, Epub 01 Jul 24.
51. Radoja N, Komine M, Jho SH et al. Novel mechanism of steroid action in skin through Glucocorticoid receptor monomers. Mol Cell Biol 2000; 20(12):4328-39.
52. Tomic-Canic M, Freedberg IM, Blumenberg M. Codominant regulation of keratin gene expression by cell surface receptors and nuclear receptors. Exp Cell Res 1996; 224(1):96-102.
53. Jiang CK, Flanagan S, Ohtsuki M et al. Disease-activated transcription factor: Allergic reactions in human skin cause nuclear translocation of stat-91 and induce synthesis of keratin K17 molecular effects of T lymphocytes on the regulation of keratin gene expression. A cluster of five nuclear proteins regulates keratin gene transcription. Mol Cell Biol 1994; 14(7):4759-69.
54. Komine M, Freedberg IM, Blumenberg M. Regulation of epidermal expression of keratin K17 in inflammatory skin diseases. J Invest Dermatol 1996; 107(4):569-75.
55. Milisavljevic V, Freedberg IM, Blumenberg M. Characterization of nuclear protein binding sites in the promoter of keratin K17 gene. DNA Cell Biol 1996; 15(1):65-74.
56. Bernerd F, Del Bino S, Asselineau D. Regulation of keratin expression by ultraviolet radiation: Differential and specific effects of Ultraviolet B and Ultraviolet A exposure. J Invest Dermatol 2001; 117(6):1421-9.
57. Jiang CK, Tomic-Canic M, Lucas DJ et al. TGF beta promotes the basal phenotype of Epidermal keratinocytes: Transcriptional induction of K#5 and K#14 keratin genes. Growth Factors 1995; 12(2):87-97.
58. Mahony D, Karunaratne S, Cam G et al. Analysis of mouse keratin 6a regulatory sequences in transgenic mice reveals constitutive, Tissue-specific expression by a keratin 6a minigene. J Invest Dermatol 2000; 115(5):795-804.
59. Mazzalupo S, Coulombe PA. A reporter transgene based on a human keratin 6 gene promoter is specifically expressed in the periderm of mouse embryos. Mech Dev 2001; 100(1):65-9.
60. Paladini RD, Coulombe PA. Directed expression of keratin 16 to the progenitor basal cells of transgenic mouse skin delays skin maturation. J Cell Biol 1998; 142(4):1035-51.
61. Seitz CS, Lin Q, Deng H et al. Alterations in NF-kappaB function in transgenic Epithelial tissue demonstrate a growth inhibitory role for NF-kappaB. Proc Natl Acad Sci USA 1998; 95(5):2307-12.

62. Rodriguez-Villanueva J, Greenhalgh D, Wang XJ et al. Human keratin-1.Bcl-2 transgenic mice aberrantly express keratin 6, exhibit reduced sensitivity to keratinocyte cell death induction, and are susceptible to skin tumor formation. Oncogene 1998; 16(7):853-63.
63. Magnaldo T, Bernerd F, Freedberg IM et al. Transcriptional regulators of expression of K#16, The disease-associated keratin. DNA Cell Biol 1993; 12:911-23.
64. Bernerd F, Magnaldo T, Freedberg IM et al. Expression of the carcinoma-associated keratin K6 and the role of AP-1 proto-oncoproteins. Gene Expr 1993; 3:187-99.
65. Komine M, Rao LS, Kaneko T et al. Inflammatory versus proliferative processes in epidermis. Tumor necrosis factor alpha induces K6b keratin synthesis through a transcriptional complex containing NF-kappaB and C/EBPbeta. J Biol Chem 2000; 275(41):32077-88.
66. Komine M, Rao LS, Freedberg IM et al. Interleukin-1 induces transcription of keratin K6 in human epidermal keratinocytes. J Invest Dermatol 2001; 116(2):330-8.
67. Jiang CK, Magnaldo T, Ohtsuki M et al. Epidermal growth factor and transforming growth factor alpha specifically induce the activation- and Hyperproliferation- associated keratins 6 and 16. Proc Natl Acad Sci USA 1993; 90:6786-90.
68. Smith FJ, Porter RM, Corden LD et al. Cloning of human, murine, and marsupial keratin 7 and a survey of K7 expression in the mouse. Biochem Biophys Res Commun 2002; 297(4):818-27.
69. Kulesh DA, Cecena G, Darmon YM et al. Posttranslational regulation of keratins: Degradation of mouse and human keratins 18 and 8. Mol Cell Biol 1989; 9:1553-65.
70. Oshima RG, Baribault H, Caulin C et al. Oncogenic regulation and function of keratins 8 and 18 identification of the gene coding for the Endo B murine cytokeratin and its methylated, stable inactive state in mouse nonepithelial cells. Cancer Metastasis Rev 1996; 15(4):445-71.
71. Mukhopadhyay T, Roth JA. P53 involvement in activation of the cytokeratin 8 gene in tumor cell lines. Anticancer Res 1996; 16(1):105-12.
72. Takemoto Y, Fujimura Y, Matsumoto M et al. The promoter of the endo A cytokeratin gene is activated by a 3' downstream enhancer. Nucl Acid Res 1991; 19:2761-65.
73. Prochasson P, Gunther M, Laithier M et al. Transcriptional mechanisms responsible for the overexpression of the keratin 18 gene in cells of a human colon carcinoma cell line. Exp Cell Res 1999; 248(1):243-59.
74. Gunther M, Frebourg T, Laithier M et al. An Sp1 binding site and the minimal promoter contribute to overexpression of the cytokeratin 18 gene in tumorigenic clones relative to that in nontumorigenic clones of a human carcinoma cell line. Mol Cell Biol 1995; 15(5):2490-9.
75. Prochasson P, Delouis C, Brison O. Transcriptional deregulation of the keratin 18 gene in human colon carcinoma cells results from an altered acetylation mechanism. Nucleic Acids Res 2002; 30(15):3312-22.
76. Rhodes K, Oshima RG. A regulatory element of the human keratin 18 gene with AP-1-dependent promoter activity. J Biol Chem 1998; 273(41):26534-42.
77. Toietta G, Koehler DR, Finegold MJ et al. Reduced inflammation and improved airway expression using helper-dependent adenoviral vectors with a K18 promoter. Mol Ther 2003; 7(5 Pt 1):649-58.
78. Pankov R, Neznanov N, Umezawa A et al. AP-1, ets, and transcriptional silencers regulate retinoic acid-dependent induction of keratin 18 in Embryonic cells. Mol Cell Biol 1994; 14(12):7744-57.
79. Umezawa A, Yamamoto H, Rhodes K et al. Methylation of an ets site in the intron enhancer of the keratin 18 gene participates in tissue-specific repression AP-1, ets, and transcriptional silencers regulate retinoic acid-dependent induction of keratin 18 in embryonic cells. Mol Cell Biol 1997; 17(9):4885-94.
80. Neznanov NS, Oshima RG. Cis regulation of the keratin 18 gene in transgenic mice. Mol Cell Biol 1993; 13(3):1815-23.
81. Neznanov N, Umezawa A, Oshima RG et al. A regulatory element within a coding exon modulates keratin 18 gene expression in transgenic mice methylation of an ets site in the intron enhancer of the keratin 18 gene participates in tissue-specific repression AP-1, ets, and transcriptional silencers regulate retinoic acid-dependent induction of keratin 18 in embryonic cells. J Biol Chem 1997; 272(44):27549-57.
82. Li Q, Stamatoyannopoulos G. Hypersensitive site 5 of the human beta locus control region functions as a Chromatin insulator. Blood 1994; 84(5):1399-401.
83. Neznanov N, Kohwi-Shigematsu T, Oshima RG et al. Contrasting effects of the Satb1 core nuclear matrix attachment region and flanking sequences of the keratin 18 gene in transgenic mice Cis regulation of the keratin 18 gene in transgenic mice. Mol Biol Cell 1996; 7(4):541-52.
84. Thorey IS, Cecena G, Reynolds W et al. Alu sequence involvement in transcriptional insulation of the keratin 18 gene in transgenic mice. Mol Cell Biol 1993; 13(11):6742-51.
85. Willoughby DA, Vilalta A, Oshima RG. An Alu element from the K18 gene confers position-independent expression in transgenic mice. J Biol Chem 2000; 275(2):759-68.

86. Brembeck FH, Rustgi AK. The tissue-dependent keratin 19 gene transcription is regulated by Gklf/Klf4 and Sp1. J Biol Chem 2000; 275(36):28230-9.
87. Hu L, Gudas LJ. Activation of keratin 19 gene expression by a 3' enhancer containing an AP1 site. J Biol Chem 1994; 269(1):183-91.
88. Hu L, Crowe DL, Rheinwald JG et al. Abnormal expression of retinoic acid receptors and keratin 19 by human oral and epidermal squamous cell carcinoma cell lines. Cancer Res 1991; 51(15):3972-81.
89. Choi I, Gudas LJ, Katzenellenbogen BS. Regulation of keratin 19 gene expression by estrogen in human breast cancer cells and identification of the estrogen responsive gene region. Mol Cell Endocrinol 2000; 164(1-2):225-37.
90. Kagaya M, Kaneko S, Ohno H et al. Cloning and characterization of the 5'-flanking region of human cytokeratin 19 gene in human cholangiocarcinoma cell line. J Hepatol 2001; 35(4):504-11.
91. Brembeck FH, Moffett J, Wang TC et al. The keratin 19 promoter is potent for cell-specific targeting of genes in transgenic mice. Gastroenterology 2001; 120(7):1720-8.
92. Oyama N, Takahashi H, Tojo M et al. Different properties of three isoforms (alpha, beta, and gamma) of transcription factor AP-2 in the expression of human keratinocyte genes. Arch Dermatol Res 2002; 294(6):273-80, Epub 2002 Jul 20.
93. Waikel RL, Wang XJ, Roop DR. Targeted expression of C-Myc in the epidermis alters normal proliferation, differentiation and UV-B induced apoptosis. Oncogene 1999; 18(34):4870-8.
94. Jaubert J, Cheng J, Segre JA. Ectopic expression of Kruppel like factor 4 (Klf4) accelerates formation of the Epidermal permeability barrier. Development 2003; 130(12):2767-77.
95. Sheng H, Goich S, Wang A et al. Dissecting the oncogenic potential of Gli2: Deletion of an Nh(2)-terminal fragment alters skin tumor phenotype. Cancer Res 2002; 62(18):5308-16.
96. Tarutani M, Cai T, Dajee M et al. Inducible activation of Ras and Raf in adult epidermis. Cancer Res 2003; 63(2):319-23.
97. Chiambaretta F, Blanchon L, Rabier B et al. Regulation of corneal keratin-12 gene expression by the human kruppel-like transcription factor 6. Invest Ophthalmol Vis Sci 2002; 43(11):3422-9.

CHAPTER 8

Simple Epithelial Keratins:
Expression, Function and Disease

M. Llanos Casanova,* Ana Bravo and José L. Jorcano

Abstract

K eratins K8 and K18 are the major components of the intermediate-filament cytoskeleton of simple epithelia. They are mostly expressed in internal one-layered epithelia.
K8 and K18 are considered the ancestral precursor genes for the multiple specialized
type II and type I keratin classes, respectively, and constitute the first keratin genes to be
expressed in the embryo. Herein we focused on the current knowledge of the functions
ascribed to simple epithelial keratins and the diseases associated with them. Because K8/K18
expression is naturally associated with cells with a great proliferation potential, such as cells
of embryonic structures and cells in the later stages of cancer, we centered on the relationship
between altered expression of K8/K18 and development and progression of tumors.

Introduction

Cytoskeleton refers to the major fibrillar elements that are found in cells and comprise
three major protein families: microfilaments, microtubules and intermediate filaments (IF),
which in turn interact with a growing list of associated proteins. Keratins are the most diverse subfamily of IF proteins, and are encoded by over 20 genes.[1,2] They are subdivided into
type I (K9-K22) and type II (K1-K8), which are coordinately expressed in specific type
I-type II pairs in epithelia.[1,2] The pattern and quantity of keratin expression is strictly regulated in epithelial cells.

Simple epithelial keratins are mostly expressed in internal one-layered epithelia. K8 and
K18 constitute a hallmark for all simple epithelial cells[1,3,4] although some of them express two
to three additional keratins (K7, K19, K20) as their differentiation progresses. Therefore, K7 is
expressed in many gland ducts and internal epithelia;[5] K19 is an unusual keratin, lacking a
proper tail domain expressed in the intestine, kidney collecting ducts, gallbladder, mesothelium, and
secretory glands;[1] and K20 is expressed in gastrointestinal epithelia, urothelium and neuroen-
docrine cells.[6] An additional potential simple epithelial keratin ("K23") was recently reported
in pancreatic epithelial cells.[7]

In humans, type I keratin genes are, all but one, located on chromosome 17, and type II
keratin genes on chromosome 12. The coding K18 gene is the only keratin type I gene located
on chromosome 12, close to the K8 gene. This observed juxtaposition supports the notion that
K8 and K18 may have an ancient and common origin and may be the ancestral precursor genes
for the multiple specialized type II and type I keratin classes, respectively.[8] The gene cluster on
chromosome 17 probable arose by a duplication and transposition of an ancestral K18 gene,

*Corresponding Author: M. Llanos Casanova—Molecular and Cell Biology, CIEMAT, Av.
Complutense 22, 28040 Madrid, Spain. Email: llanos.casanova@ciemat.es

Intermediate Filaments, edited by Jesus Paramio. ©2006 Landes Bioscience
and Springer Science+Business Media.

followed by concerted duplication of keratin genes in both locations. The driving force for the diversification to the obligate heteropolymeric keratin groups is not clear.

K8/K18 constitute the first keratin genes to be expressed in the embryo[9-11,] starting at the eight-cell stage of the mouse embryo, followed by K19 and K7. Probably, this very early expression was the reason for their high fidelity conservation across a wide range of species through evolution and for the apparent absence of K8/K18 mutations in the positions associated with severe disease in stratified epithelial keratins -the end domains of the α-helical rod domains-, suggesting a lethal phenotype of such mutations[12] in simple epithelial keratins.

Suggested Functions of Simple Epithelial Keratins

The study of several transgenic mouse lines targeting simple epithelial keratins has been very useful to elucidate the requirement of keratins for the formation and function of simple epithelia. They include the targeted null mutations of K8, K18 and K19. Because of their early expression it has been assumed that these keratins have an important function during embryogenesis. The first reported IF knockout were the K8-null mice that have variable phenotypes depending on their genetic background. In one strain (C57Bl/6), the targeted null mutation of K8 causes mid-gestational lethality with placental bleeding due to TNF-sensitive failure of trophoblast giant cell barrier function.[13,14] Approximately half of the homozygous null embryos escape lethality in mice with the FVB/n genetic background. These K8 deficient mice have colorectal hyperplasia and inflammation, rectal prolapse, female sterility, and, in some cases, mild hepatic coagulative necrosis.[15] In contrast, K18 null mice, which also lack hepatocyte keratin filaments, are viable and fertile and their liver morphology appears relatively normal, apart from accumulation of K8-containing Mallory bodies and formation of enlarged hepatocytes in aging mice.[16] More recently, additional detailed abnormal liver and pancreas phenotype were observed in both K8 and K18 null mice. They consisted, mainly of large polynuclear areas that lacked cell membranes, desmosomal structures, and filamentous actin.[17] The finding that K8- and K18- null mice were viable was unexpected and it was hypothesized that K19 and K7, which begin to be expressed shortly after K8/K18, could compensate for the loss of K18 and K8 respectively. K19 null mice were generated and proved to be normal.[18] In contrast, the generation of doubly deficient K18/K19 mice[19] in which all embryonic type I keratins have been abolished causes the death of fetuses at embryonic day 9.5 due to severe cytolysis in trophoblast giant cells. And doubly deficient K8/K19 mice[18] result in mice that suffer early embryonic lethality around embryonic day 10 due to anomalies in the trophoblast layer along with major placental defects, with 100% penetrance. These doubly knockout mice have served to demonstrate a primary cytoskeletal function of these keratins, indicating that they have an essential function during embryonic development, at least in the extraembryonic compartments of the conceptus. Conceivably, an approach that includes the aggregation of null embryos with tetraploid wild-type embryos would help to distinguish between the function of simple epithelial keratin in embryos as such from their role in extraembryonic tissues.

The Structural Function of Simple Epithelial Keratins

Important information regarding keratin function was also obtained by overexpression of wild type keratins or the expression of mutant or ectopic keratins. While the role of epidermal keratins in the maintenance of the structural integrity of the epidermis is widely accepted,[20-22] the structural function of simple epithelial keratins is less clear. A mouse model of transgenic mice with disrupted K8/K18 filament assembly originated by expression of a K18 containing a mutated arg89→cys (a highly conserved site among keratins and a frequently mutated amino acid in several human skin diseases[2]), develop mild chronic hepatitis and markedly fragile hepatocytes in association with K8/18 filament disruption.[23] In addition, these mice as well as K8 null mice were more susceptible to hepatotoxic damage.[24,25] These results indicate that one function of keratins in hepatocytes is to maintain cellular integrity and resistance and that an intact keratin filament network is important for protection from drug-induced liver injury (the roles of keratins in hepatic function is further disccussed in the chapter by Zatloukal et al).

However, the cell-protective role that K8/K18 filaments impart in the liver was not observed in the pancreas.[26,27] Therefore, differences in hepatocyte and acinar cell susceptibility to injury, despite their similarity in terms of cytoplasmic filament disruption, raise the possibility that identical keratins may function differently in different cells.

Involvement of K8/K18 in the Maintenance of the Cell Polarity and Tissue Architecture of Simple Epithelial Cells

Another role proposed for simple epithelial keratins is their involvement in the maintenance of cell polarity. It is known that several cell types develop a polarized structure that is essential for their biological function; typically polarized cells are neurons and epithelial cells. Cell polarization is especially important in simple epithelium due to its secretory and absorptive functions. Keratin distribution is often asymmetric in simple epithelial cells, with filaments concentrated toward the apical surface (Fig. 1). There is growing evidence that keratins are involved in the maintenance of the cell polarity and tissue architecture of simple epithelial cells. Therefore, enterocytes lose the polarity of different components when the K19 (usually concentrated in the apical cortical region of the cell) is depleted.[28] Villus enterocytes of K8 null mice lose keratins and apical membrane markers and undergo disorganization of microtubules.[29] We have reported that acinar cells of the exocrine pancreas of transgenic mice expressing the wild type human K8 (TGHK8 mice) and with just a three-fold increase in the amount of K8/K18, lost K8/K18 keratin filaments polarization. These appear distributed throughout the whole cytoplasm instead of concentrated at the apical region of the cell, where they are localized in nontransgenic acinar cells[30] (Fig. 1). Histologically, TGHK8 mice show a marked disorganization of the exocrine pancreatic tissue structure with loss of the typical organization of the secretory units arranged in acini, with the apical region of acinar cells oriented toward the central lumen (Fig. 1). Recent studies revealed an additional abnormal liver phenotype (also affecting, to a lesser extent, the exocrine pancreas) in K8- and K18 null mice, and in K18 arg→cys mice, related to a complete lack of organization in some parts of hepatic tissue and a disturbed cellular organization and turnover in the rest of the tissue.[17] These results strengthen the role of simple epithelial keratins in maintaining the architecture of the tissues that normally express them.

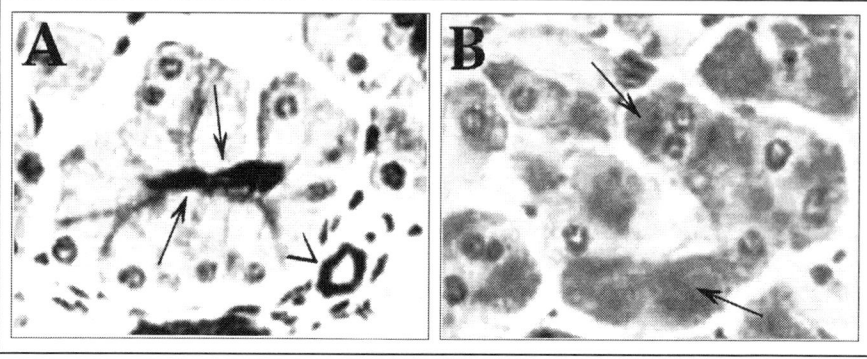

Figure 1. Immunodetection of K8 in pancreas sections. Paraffin-embedded section from wild type (A) and transgenic TGHK8 (B) mice. Staining was performed with TROMA 1[71] antibody which recognizes both, the endogenous mouse K8 and the human K8. In wild type pancreas the signal is restricted to the apical region of acinar cells (arrows) and to ductal cells (arrowhead). In transgenic pancreas sections the staining appears throughout the cytoplasm of acinar cells (arrows). Note the disorganization of the pancreatic tissue structure in transgenic mice.

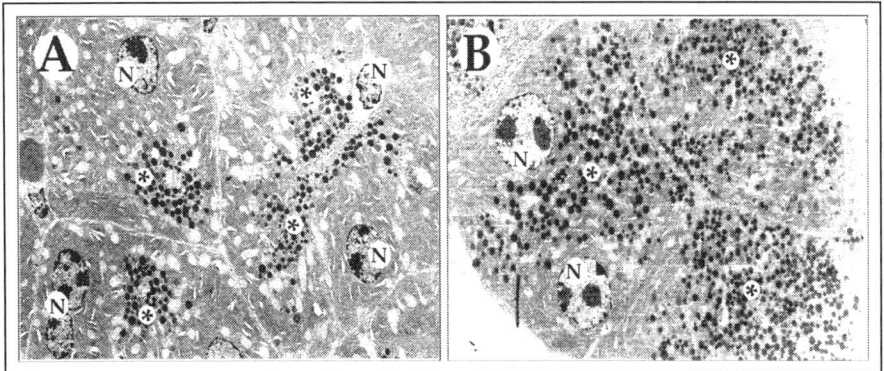

Figure 2. Ultrastructural analysis of pancreatic acinar cells. A) Electron micrograph showing the apical localization of zymogen granules in wild type cells. B) Note the accumulation and loss of the apical localization of zymogen granules in transgenic TGHK8 cells. N, nucleus; asterisks, zymogen granules. 1500X.

Finally, associated with the role of K8/K18 in maintaining cell polarity there is growing evidence for a role of simple epithelial keratins in vesicular trafficking. For instance TGHK8 mice present striking exocrine enzymatic pancreatic deficiencies,[30] possibly as a consequence of impaired transport of zymogen granules to the apical region of acinar cells (Fig. 2). Also, abnormalities in the distribution of apical surface markers in both the small intestine and liver of K8 null animals were identified. Regional differences were found in the expression of syntaxin-3, a protein involved in vesicular transport.[29] The mechanism by which keratin filaments might influence vesicular trafficking remains to be elucidated.

Possible Implication of K8/K18 in Signaling

The phenotype of the TGHK8 mouse that displays loss of cellular architecture, acino-ductular metaplasia, fibrosis and adipose replacement of wide regions of the exocrine pancreas[30] is very similar to that of the transgenic mice expressing a dominant negative mutant TGF-β type II receptor.[31] Accordingly, the latter also show increased levels of K8 and K18 in their pancreatic acinar cells, suggesting the involvement of these keratins in TGF-β signaling.[30] Other experimental findings have also suggested the involvement of simple epithelial keratins in signaling, as shown by their interaction with different cytoplasmic proteins. In particular, the interaction of K18 with 14-3-3 proteins in a phosphorylation-dependent manner[32,33] appear to partially modulate hepatocyte mitotic progression in association with alterations in the nuclear distribution of 14-3-3 proteins during mitosis.[34] In agreement, the distribution of the 14-3-3ζ protein in the liver of K8 null mice (lacking keratin K8/K18 filaments in hepatocytes) is different to that seen in wild-type animals.[17] Keratins have also been reported to associate with kinases, including a fragment of the PKC-epsilon[35] and JNK,[36] yielding in this case, the keratin 8 as a cytoplasmic target for JNK in Fas receptor-mediating signaling (see implication of K8/K18 in apoptotic signaling). More recently, the defective formation of desmosomes by assembly deficiency in K8/K18 filaments[17] has been described. Because desmosomes contain both plakoglobin—a protein involved in developmental signaling—and cadherin family members, their function may be more than a strictly mechanical adhesion between cells. They could also participate in mediating cell-cell signaling transduction, suggesting an important role for simple epithelial keratins in signal transduction.

Keratins K8/K18 May Be Involved in Apoptotic Signaling

The association of keratins to apoptosis occurs at several levels. On one hand the type I simple epithelial keratins are early targets of caspase activity in apoptosis. Cleavage of one keratin will destabilize the filament network due to the heteropolymeric nature of these proteins. The fragmentation of keratins by caspases may facilitate apoptotic cell clearing by allowing the keratin network to be dismantled.[37] On the other hand, several studies have revealed a protective role for keratin 8 during TNF and Fas mediated apoptosis. Therefore, it has been described that the K8/K18 pair can desensitize cells to pro-apoptotic signaling mediated by tumor necrosis-alpha (TNF-α)[38] or by Fas ligand,[39] by binding to their receptors. K18 also binds a downstream TNF receptor, TRADD.[40]

TNF-α is an inflammatory cytokine that is elevated in inflammatory bowel disorders and is also involved in liver apoptosis. Therefore, the inhibition of pro-apoptotic signaling by simple epithelial keratins would help to explain some observations performed in K8- and K18- null mice (lacking all keratin IF in their hepatocytes). These mice are very sensitive to Fas-mediated apoptosis or to concanavalin A-induced apoptosis, which is mediated by Fas and TNF.[39,41] This could also explain the sensitivity of K8 and K18 null hepatocytes to some toxins. In addition, the association of K8/K18 with death receptor family function may be related to an earlier association of K8/K18 with drug resistance in several cell lines.[42,43] Some K8/K18 deficient cell lines were shown to be more sensitive to DNA damaging agents than the controls that expressed both keratins. These findings also support the involvement of K8/K18 mutations in gastrointestinal diseases. It has been reported that stimulation of the Fas receptor of intestinal epithelial cells induces activation of the stress kinase JNK and phosphorylation of K8 in serine-73.[36] This coincides with the association of some JNK with K8 and a decreased ability of JNK to phosphorylate c-jun. Thus, the presence or absence of K8/K18 could modulate signaling from Fas at this stage in the pathway, as well as by interfering with receptor targeting.

Finally, very recently, the lethality of K8 null embryos has been reported to be due to a TNF-sensitive failure of trophoblast giant cell barrier function.[14] This will present the keratin-dependent protection of trophoblast giant cells from a maternal TNF-dependent apoptotic challenge as a key role of simple epithelial keratins.

The Tissue-Specific Expression of Keratins Implies Unique Functions

Finally, the tissue- and cell type-specific distribution of keratins in epithelial cells suggests a unique functional role of these proteins in these cell types. Keratin replacement experiments in transgenic mice reveal that keratins are not completely redundant. For example, introduction of K18 into transgenic mice lacking K14, by using the K14 promoter, only partially rescues the lethal and blistering phenotype typically displayed by K14-null mice.[44] This suggests that different keratins are likely to serve tissue-specific roles that reflect their unique expression.

Simple Epithelial Keratin Proteins and Disease

It is difficult to associate mutations with disorders of simple epithelia. They may be expressed in vital organs and at early stages of development that would cause embryonic lethality. Futhermore, the structural role of keratins may be less central to the function of keratins in simple epithelia than in the epidermis, because internal tissues are normally not subjected to stresses and therefore the structural diseases caused by mutations of epidermal keratins might not occur in simple epithelia. Within this context, the study of animal models carrying null or dominant negative mutations or overexpression of simple epithelial keratins suggest that in addition to carrying a mutation in a keratin, the simple epithelial tissue might have to be abnormally stressed (e.g., chemical toxicity, infection or inflammation) for a phenotype to manifest itself.[24,25] Authors seems to agree that K8 or K18 mutations predispose, rather than cause, their carriers to develop a disease.[45,46]

The K8/K18 transgenic animal models suggested that the liver, the intestine and the pancreas are the likely target of K8/K18 mutations if they are to occur in humans. The most likely population of K8/K18 mutations is patients with acute or chronic cryptogenic liver disease.[12,47,48] In addition, an intestinal target, as in inflammatory bowel disease, may be associated with K8 mutations, based on the intestinal phenotype of the K8-null mice[15] and on the increased sensitivity to TNF-α liver mediated apoptosis observed in these mice.[14] Very recently it has been found that the glycine to cysteine mutation at position 61 of the keratin 8 gene, together with other enviromental factors and/or genetic factors, could predispose to chronic pancreatitis, by interfering with the normal organization of keratin filaments.[49] Simple epithelial keratin diseases would be rare. However, if the predisposition model were correct, it could potentially explain a greater incidence of association between simple keratin disorders and epithelial disease.

Altered Simple Epithelial Keratin Expression and Cancer

The detection of K8/K18 is used for the diagnosis and prognosis of tumors of epithelial and nonepithelial[50,51] origin. Proteolytic fragments of K8 and K18 (most likely through apoptotic cleavage) are components of the tissue polypeptide antigen, found in large amounts in the serum of patients with a wide range of carcinomas, and their concentration correlates with survival of these patients.[50,51] However, beyond the diagnostic use of K8 and K18, the functional significance of this correlation remains unclear and no role for K8/K18 in tumorigenesis has yet been found.

It has been observed that some tumors derived from simple epithelia retain and augment K8 and K18 expression. This is the case of different grade and stage of transitional cell carcinomas (TCC) of the urinary tract.[52] An increase in K8/K18 expression has also been detected concomitant with an increase in the degree of malignancy of the tumor as in the case of the increased expression of K8/K18 in the most malignant infiltrating TCCs.[53] A possible role for K8 and K18 in increasing invasiveness and migration properties of tumor cells has also been suggested by transfection experiments in different tumor cells.[54-57]

On the other hand, albeit in adults, K8 and K18 constitute the major intermediate filaments in simple epithelial tissues such as liver, intestine and pancreas, but are not normally present in the stratified epithelia such as epidermis. A well-known but unexplained clinical phenomenon is the aberrant expression of simple epithelial keratins by different types of human carcinomas of stratified epithelial origin, e.g., oral and skin cancers. A positive correlation has been established between the presence of these keratins and tumor malignancy both in vitro[58] and in vivo.[59-61] Even more intriguing is the aberrant expression of K8 and K18 in human tumor cells of different nonepithelial origins, such as malignant melanoma and lymphoma, which normally only express type-III IF vimentin.[55,62]

It has been reported that K8 expression is induced by v-*ras* in transformed epidermal keratinocytes.[63,64] Also, the K8 gene is induced by EGFR activation,[65] EGFR being overexpressed in both human and mouse epidermal tumors.[66,67] Conversely, abrogation of EGFR function inhibits tumor development.[68,69] The important role of ras and EGFR in the induction of epidermal tumors and their progression would suggest that transcription factors activated by the ras and EGFR signaling transduction pathway may mediate the aberrant expression of K8 and K18. To determine whether a causal relationship exists between the abnormal expression of simple epithelium keratins and the development of neoplastic characteristics and, given the increasing incidence of epidermal cancer in humans (40% of all tumors diagnosed[70]), we have generated a transgenic mouse model (TGHK8 mice) to examine the role of simple epithelium keratins in the establishment and progression of human skin cancer (Casanova et al submitted results). Skin carcinogenesis assays performed in our TGHK8 mice expressing the human K8 in the epidermis showed a dramatic increase in the progression of papillomas towards malignancy in transgenic animals. Thus, our results would support the notion that K8/K18 expression is not maintained in carcinomas because the regulation of K8/K18 genes uses oncogene activated transcription factors. Further, our findings demonstrate that K8 expression contributes directly to the malignant skin phenotype, and that this keratin may function as a promoter of malignancy.

The identification of the possible alterations in signaling molecules and pathways that lead to the malignization of these tumors—that in turn exhibit an aberrant expression of K8, will contribute to the understanding of pathogenic states of the skin associated with epidermal hyperplasia and dysplastic changes and, may ultimately lead to the development of new therapies for skin tumors.

Concluding Remarks

The different studies of keratin functions and disorders allow authors to conclude that a function common to all keratins is the structural function. It also appears that this is not the main function of keratins in simple epithelia. For this reason, the structural diseases caused by mutations of epidermal keratins might not occur in simple epithelia. Another interesting finding is that in simple epithelial cells the same keratins may display different functions in different cell types. The recent suggestions of the involvement of simple epithelial keratins in signal transduction pathways indicate the relevance of maintaining the tissue-specific and amount of keratin expression in a cell for the appropriate development of its physiological functions. In addition, the role of simple epithelial keratins as inhibitors of pro-apoptotic signaling may be essential in those stages where a fine regulation of proliferation/death is necessary as in the case of embryogenesis and cancer. A very important issue is the increasing evidence that points to the aberrant increased expression levels of simple epithelial keratins in a transformed cell as a promoter of an invasive behavior of tumors of different origin. This makes the study of the expression of these proteins in cancer a valuable indicator of tumor evolution.

Note Added in Proof

The data referred to as "Casanova et al, submitted results" have been published (Epidermal abnormalities and increased malignancy of skin tumors in human epidermal keratin 8-expressing transgenic mice. FASEB J 2004, Aug. 19). In addition, during the revision of the manuscript it has been published in a work by Vaidya's group reinforcing the induction of transformed phenotype and invasiveness by forced K8 expression in stratified epithelial cells (Implications of cytokeratin 8/18 filament formation in stratified epithelial cells: Induction of cytokeratin phenotype. Raul et al. Int J Cancer 2004; 111:662-658).

References

1. Moll R, Franke WW, Schiller D. The catalog of human cytokeratins: Patterns of expression in normal epithelia, tumors and cultured cells. Cell 1982; 31(1):11-24.
2. Fuchs E, Cleveland DW. A structural scaffolding of intermediate filaments in health and disease. Science 1998; 279(5350):514-519.
3. Oshima RG, Baribault H, Caulin C. Oncogenic regulation and function of keratins 8 and 18. Cancer Metastasis Rev 1996; 15(4):445-471.
4. Omary MB, Ku NO. Intermediate filament proteins of the liver: Emerging disease association and functions. Hepatology 1997; 25(5):1043-1048.
5. Smith FJ, Porter RM, Corden LD et al. Cloning of human, murine, and marsupial keratin 7 and a survey of K7 expression in the mouse. Biochem Biophys Res Commun 2002; 297(4):818-827.
6. Moll R, Lowe A, Laufer J et al. Cytokeratin 20 in human carcinomas. A new histodiagnostic marker detected by monoclonal antibodies. Am J Pathol 1992; 140(2):427-447.
7. Zhang JS, Wang L, Huang H et al. Keratin 23 (K23), a novel acidic keratin, is highly induced by histone deacetylase inhibitors during differentiation of pancreatic cancer cells. Genes Chromosomes Cancer 2001; 30(2):123-135.
8. Blumenberg M. Concerted gene duplications in the two keratin gene families. J Mol Evol 1988; 27(3):203-211.
9. Jackson BW, Grund C, Schmid E et al. Formation of cytoskeletal elements during mouse embryogenesis. Intermediate filaments of the cytokeratin type and desmosomes in preimplantation embryos. Differentiation1980; 17(3):161-179.
10. Oshima RG, Howe WE, Klier FG et al. Intermediate filament protein synthesis in preimplantation murine embryos. Dev Biol 1983; 99(2):447-455.

11. Lehtonen E, Lehto VP, Vartio T et al. Expression of cytokeratin polypeptides in mouse oocytes and preimplantation embryos. Dev Biol 1983; 100(1):158-165.

12. Ku NO, Darling JM, Krams SM et al. Keratin 8 and 18 mutations are risk factors for developing liver disease of multiple etiologies. Proc Natl Acad Sci USA 2003; 100(10):6063-6068.

13. Baribault H, Price J, Miyai K et al. Mid-gestational lethality in mice lacking keratin 8. Genes Dev 1993; 7(7A):1191-1202.

14. Jaquemar D, Kupriyanov S, Wankell M et al. Keratin 8 protection of placental barrier function. J Cell Biol 2003; 161(4):749-756.

15. Baribault H, Penner J, Iozzo RV et al. Colorectal hyperplasia and inflammation in keratin 8-deficient FVB/N mice. Genes Dev 1994; 8(24):2964-2973.

16. Magin TM, Schroder R, Leitgeb S et al. Lessons from keratin 18 knockout mice: Formation of novel keratin filaments, secondary loss of keratin 7 and accumulation of liver-specific keratin 8-positive aggregates. J Cell Biol 1998; 140(6):1441-1451.

17. Toivola DM, Nieminen MI, Hesse M et al. Disturbances in hepatic cell-cycle regulation in mice with assembly-deficient keratins 8/18. Hepatology 2001; 34(6):1174-1183.

18. Tamai Y, Ishikawa T, Bosl MR et al. Cytokeratins 8 and 19 in the mouse placental development. J Cell Biol 2000; 151(3):563-572.

19. Hesse M, Franz T, Tamai Y et al. Targeted deletion of keratins 18 and 19 leads to trophoblast fragility and early embryonic lethality. EMBO J 2000; 19(19):5060-5070.

20. Fuchs E, Weber K. Intermediate filaments: Structure, dynamics, function, and disease. Annu Rev Biochem 1994; 63:345-382.

21. Ma L, Yamada S, Wirtz D et al. A 'hot-spot' mutation alters the echanical properties of keratin filament networks. Nat Cell Biol 2001; 3(5):503-506.

22. Porter RM, Lane EB. Phenotypes, genotypes and their contribution to understanding keratin function. Trends Genet 2003; 19:278-285.

23. Ku NO, Michie S, Oshima RG et al. Chronic hepatitis, hepatocyte fragility, and increased soluble phosphoglycokeratins in transgenic mice expressing a keratin 18 conserved arginine mutant. J Cell Biol 1995; 131(5):1303-1314.

24. Toivola DM, Omary MB, Ku NO et al. Protein phosphatase inhibition in normal and keratin 8/18 assembly-incompetent mouse strains supports a functional role of keratin intermediate filaments in preserving hepatocyte integrity. Hepatology 1998; 28(1):116-128.

25. Zatloukal K, Stumptner C, Lehner M et al. Cytokeratin 8 protects from hepatotoxicity, and its ratio to cytokeratin 18 determines the ability of hepatocytes to form Mallory bodies. Am J Pathol 2000; 156(4):1263-1274.

26. Toivola DM, Baribault H, Magin T et al. Simple epithelial keratins are dispensable for cytoprotection in two pancreatitis models. Am J Physiol Gastrointest Liver Physiol 2000; 279(6):G1343-1354.

27. Toivola DM, Ku NO, Ghori N et al. Effects of keratin filament disruption on exocrine pancreas-stimulated secretion and susceptibility to injury. Exp Cell Res 2000; 255(2):156-170.

28. Salas PJ, Rodriguez ML, Viciana AL et al. The apical submembrane cytoskeleton participates in the organization of the apical pole in epithelial cells. J Cell Biol 1997; 137(2):359-375.

29. Ameen NA, Figueroa Y, Salas PJ. Anomalous apical plasma membrane phenotype in CK8-deficient mice indicates a novel role for intermediate filaments in the polarization of simple epithelia. J Cell Sci 2001; 114(Pt 3):563-575.

30. Casanova ML, Bravo A, Ramírez A et al. Exocrine pancreatic disorders in transgenic mice expressing human keratin 8. J Clinical Invest 1999; 103(11):1587-1595.

31. Bottinger EP, Jakubczak JL, Roberts IS et al. Expression of a dominant-negative mutant TGF-beta type II receptor in transgenic mice reveals essential roles for TGF-beta in regulation of growth and differentiation in the exocrine pancreas. EMBO J 1997; 16(10):2621-2633.

32. Liao J, Omary MB. 14-3-3 proteins associate with phosphorylated simple epithelial keratins during cell cycle progression and act as a solubility cofactor. J Cell Biol 1996; 133(2):345-357.

33. Ku NO, Liao J, Omary MB. Phosphorylation of human keratin 18 serine 33 regulates binding to 14-3-3 proteins. EMBO J 1998; 17(7):1892-1906.

34. Ku NO, Michie S, Resurreccion EZ et al. Keratin binding to 14-3-3 proteins modulates keratin filaments and hepatocyte mitotic progression. Proc Natl Acad Sci USA 2002; 99(7):4373-4378.

35. Omary MB, Baxter GT, Chou CF et al. PKC epsilon-related kinase associates with and phosphorylates cytokeratin 8 and 18. J Cell Biol 1992; 117(3):583-593.

36. He T, Stepulak A, Holmstrom TH et al. The intermediate filament protein keratin 8 is a novel cytoplasmic substrate for c-Jun N-terminal kinase. J Biol Chem 2002; 277(13):10767-10774.

37. Caulin C, Salvesen GS, Oshima RG. Caspase cleavage of keratin 18 and reorganization of intermediate filaments during epithelial cell apoptosis. J Cell Biol 1997; 138(6):1379-1394.

38. Caulin C, Ware CF, Magin TM et al. Keratin-dependent, epithelial resistance to tumor necrosis factor-induced apoptosis. J Cell Biol 2000; 149(1):17-22.
39. Gilbert S, Loranger A, Daigle N et al. Simple epithelium keratins 8 and 18 provide resistance to Fas-mediated apoptosis. The protection occurs through a receptor-targeting modulation. J Cell Biol 2001; 154(4):763-773.
40. Inada H, Izawa I, Nishizawa M et al. Keratin attenuates tumor necrosis factor-induced cytotoxicity through association with TRADD. J Cell Biol 2001; 155(3):415-426.
41. Oshima RG. Apoptosis and keratin intermediate filaments. Cell Death Differ 2002; 9(5):486-492.
42. Bauman PA, Dalton WS, Anderson JM et al. Expression of cytokeratin confers multiple drug resistance. Proc Natl Acad Sci USA 1994; 91(12):5311-5314.
43. Anderson JM, Heindl LM, Bauman PA et al. Cytokeratin expression results in a drug-resistant phenotype to six different chemotherapeutic agents. Clin Cancer Res 1996; 2(1):97-105.
44. Hutton E, Paladini RD, Yu QC et al. Functional differences between keratins of stratified and simple epithelia. J Cell Biol 1998; 143(2):487-499.
45. Owens DW, Lane EB. The quest for the function of simple epithelial keratins. Bioessays 2003; 25(8):748-758.
46. Porter RM, Lane EB. Phenotypes, genotypes and their contribution to understanding keratin function. Trends Genet 2003; 19(5):278-285.
47. Ku NO, Wright TL, Terrault NA et al. Mutation of human keratin 18 in association with cryptogenic cirrhosis. J Clin Invest 1997; 99(1):19-23.
48. Ku NO, Gish R, Wright TL et al. Keratin 8 mutations in patients with cryptogenic liver disease. N Engl J Med 2001; 344(21):1580-1587.
49. Cavestro GM, Frulloni L, Nouvenne A et al. Association of keratin 8 gene mutation with chronic pancreatitis. Dig Liver Dis 2003; 35(6):416-420.
50. Weber K, Osborn M, Moll R et al. Tissue polypeptide antigen (TPA) is related to the nonepidermal keratins 8, 18 and 19 typical of simple and nonsquamous epithelia: Reevaluation of a human tumor marker. EMBO J 1984; 3(11):2707-2714.
51. Einarsson R. TPS—a cytokeratin marker for therapy control in breast cancer. Scand. J Clin Lab Invest Suppl 1995; 221:113-115.
52. Schaafsma HE, Ramaekers FC, van Muijen GN et al. Distribution of cytokeratin polypeptides in human transitional cell carcinomas, with special emphasis on changing expression patterns during tumor progression. Am J Pathol 1990; 136(2):329-343.
53. Schaafsma HE, Van Der Velden LA, Manni JJ et al. Increased expression of cytokeratins 8, 18 and vimentin in the invasion front of mucosal squamous cell carcinoma. J Pathol 1993; 170(1):77-86.
54. Chu YW, Duffy JJ, Seftor EA et al. Transfection of a deleted CK18cDNA into a highly metastatic melanoma cell line decreases the invasive potential. Clin Biotech 1991; 3:27-33.
55. Hendrix MJ et al. Coexpression of vimentin and keratins by human melanoma tumor cells: Correlation with invasive and metastatic potential. J Natl Cancer Inst 1992; 84(3):165-174.
56. Chu YW, Runyan RB, Oshima RG et al. Expression of complete keratin filaments in mouse L cells augments cell migration and invasion. Proc Natl Acad Sci USA 1993; 90(9):4261-4265.
57. Chu YW, Seftor EA, Romer LH et al. Experimental coexpression of vimentin and keratin intermediate filaments in human melanoma cells augments motility. Am J Pathol 1996; 148(1):63-69.
58. Caulín C, Bauluz C, Gandarillas A et al. Changes in keratin expression during malignant progression of transformed mouse epidermal keratinocytes. Exp Cell Res 1993; 204(1):11-21.
59. Markey AC, Birgitte LE, Churchill LJ et al. Expression of simple epithelial keratins 8 and 18 in epidermal neoplasia. J Invest Dermatol 1991; 97:763-770.
60. Larcher F, Bauluz C, Díaz-Guerra M et al. Aberrant expression of the simple epithelial type II keratin 8 by mouse skin carcinomas but not papillomas. Mol Carcinog 1992; 6(2):112-121.
61. Ogden GR, Cowpe JG, Chisholm DM et al. DNA and keratin analysis of oral exfoliative cytology in the detection of oral cancer. Eur J Cancer B Oral Oncol 1994; 30B:405-408.
62. Lasota J, Hyjek E, Koo CH et al. Cytokeratin-positive large-cell lymphomas of B-cell lineage. A study of five phenotypically unusual cases verified by polymerase chain reaction. Am J Surg Pathol 1996; 20(3):346-354.
63. Cheng C, Kilkenny A, Roop D et al. The v-ras oncogene inhibits the expression of differentiation markers and facilitates expression of cytokeratins 8 and 18 in mouse keratinocytes. Mol Carcinog 1990; 3(6):363-373.
64. Díaz-Guerra M, Haddow S, Bauluz C et al. Expression of simple epithelial cytokeratins in mouse epidemal keratinocytes harboring Harvey ras gene alterations. Cancer Res 1992; 52(3):680-687.
65. Dlugosz A A, Cheng C, Denning MF et al. Keratinocyte growth factor receptor ligands induce transforming growth factor alpha expression and activate the epidermal growth factor receptor signaling pathway in cultured epidermal keratinocytes. Cell Growth Differ 1994; 5(12):1283-1292.

66. Hunts J, Ueda M, Ozawa S et al. Hyperproduction and gene amplifications of the EGFR in squamous cell carcinomas. Jpn J Cancer Res 1985; 76(8):633-666.

67. Rho O, Beltran LM, Conti-Giménez IB et al. Altered expression of the epidermal growth factor receptor and transforming growth factor-alpha during multistage skin carcinogenesis in SENCAR mice. Mol Carcinog 1994; 11(1):19-28.

68. Sibilia M, Fleischmann A, Behrens A et al. The EGF receptor provides an essential survival signal for SOS-dependent skin tumor development. Cell 2000; 102(2):211-220.

69. Casanova ML, Larcher F, Casanova B et al. A critical role for *ras*-mediated, EGFR-dependent angiogenesis in mouse skin carcinogenesis. Cancer Res 2002; 62(12):3402-3407.

70. Limmer BL. Nonmelanoma skin cancer: Today's epidemic. Tex Med 2001; 97(2):56-58.

71. Brulet P, Babinet C, Kemler R et al. Monoclonal antibodies against trophectoderm-specific markers during mouse blastocyst formation. Proc Natl Acad Sci USA 1980; 77(7):4113-4117.

CHAPTER 9

Keratins as Targets in and Modulators of Liver Diseases

Kurt Zatloukal,* Conny Stumptner, Andrea Fuchsbichler and Helmut Denk

Abstract

The keratin cytoskeleton of hepatocytes is affected in a variety of chronic liver diseases, such as alcoholic and nonalcoholic steatohepatitis (ASH, NASH), copper toxicosis, cholestasis and hepatocellular carcinoma. In these diseases hepatocytes reveal a derangement or even loss of the cytoplasmic keratin intermediate filament cytoskeleton and formation of cytoplasmic inclusions (Mallory bodies) consisting of misfolded and aggregated keratin as well as a variety of stress proteins. Keratin gene knock-out mice demonstrated that keratins fulfil besides a structural role providing mechanical stability to hepatocytes a role as target and modulator of toxic stress responses in that keratins interact with a variety of stress-related signaling pathways, are preferred targets of stress-induced protein misfolding and are substrates for caspases. Furthermore, the identification of mutations in keratin genes in patients with liver cirrhosis suggests that keratins act as genetic modifiers in liver diseases.

The Keratin Cytoskeleton of Hepatocytes and Bile Duct Epithelial Cells

The keratin intermediate filament (IF) cytoskeleton of hepatocytes is composed of the type I keratin 18 (K18) and type II keratin 8 (K8) in equimolar ratios.[1-3] Bile duct epithelia have a more complex keratin composition and express in addition to K8 and K18 also K7 and K19. Since K7 is present in higher concentrations than K19 the major partner for K7 in bile duct epithelia is K18. This preferred partnership of K7 and K18 was demonstrated in K18 knock-out mice where in the absence of K18 there was also a loss of K7 in bile duct epithelia.[4] In certain liver diseases, such as alcoholic steatohepatitis (ASH) and chronic cholestasis also single hepatocytes as well as small groups of hepatocytes express in addition to K8 and K18, K7 and, to a lesser extent, K19. This phenomenon was originally regarded as evidence for ductular metaplasia (i.e., hepatocytes acquire features of bile duct epithelial).[5,6] A metaplastic transition of hepatocytes to bile duct epithelia, however, was so far not observed in mouse models for ASH or cholestasis.[7] Another explanation for the occurrence of K7-positive hepatocytes is that in certain liver diseases regeneration of hepatocytes involves proliferation and differentiation of hepatocytic precursor cells (oval cells), which normally express K8, K18 as well as K7 and K19.[8,9] Therefore, the presence of hepatocytes expressing K7 indicates that these hepatocytes originated from precursor cells and not by proliferation of differentiated hepatocytes, the latter of which is the classical mode of hepatocyte regeneration.[9]

*Corresponding Author: Kurt Zatloukal—Institute of Pathology, Medical University of Graz, A-8036 Graz, Austria. Email: kurt.zatloukal@meduni-graz.at

Intermediate Filaments, edited by Jesus Paramio. ©2006 Landes Bioscience and Springer Science+Business Media.

The biological significance of the expression of different keratin pairs in bile duct epithelia and the expression of bile duct-type keratins in hepatocytes is unclear. Studies in keratin gene knock-out mice demonstrated that a regular liver can be formed in the absence either of K8, K18 or K19.[4,10-12] However mice lacking K8 or K18 or expressing a mutated K18 gene were more sensitive to a variety of toxins and stress conditions.[13] Furthermore, studies performed with epidermal keratins provided evidence that different keratin family members are associated with distinct cellular functions. In the epidermis keratinocytes replace the keratin pairs K5 / K14 by K1 / K10 during their course of differentiation.[14,15] In situations of wound healing and in neoplasia keratinocytes start expressing different keratins, such as K16, which is associated with enhanced cell proliferation and migration.[16,17] A relationship between specific keratin proteins and cell function was recently demonstrated in transfected keratinocytes, where expression of K16 resulted in higher cell proliferation rates whereas expression of K10 reduced proliferation.[18,19]

Alterations of the Keratin Cytoskeleton in Human Liver Diseases

Alterations of the hepatocytic keratin IFs is a characteristic feature of ASH, which develops in approximately 20% of heavy drinkers and rapidly leads to the development of liver cirrhosis the major cause of death of alcoholics. In ASH hepatocytes on the one hand accumulate fat in the cytoplasm (steatosis), because of a general disturbance of lipid metabolism and, on the other hand, there are non steatotic hepatocytes, which are enlarged (ballooned) and may contain cytoplasmic protein aggregates termed Mallory bodies (MBs). Further morphologic alterations in ASH are a chicken wirelike fibrosis, an inflammatory reaction with predominantly polymorphonuclear granulocytes, and cholestasis.[20-22] It has been demonstrated more than 20 years ago that the major cellar structure that is affected in ballooned hepatocytes is the keratin cytoskeleton. Ballooned hepatocytes reveal a derangement or even loss of the cytoplasmic keratin IFs and aggregation of keratin proteins as MBs[23] (Fig. 1). These alterations of the keratin cytoskeleton are characteristic but not specific of ASH since they are also seen in a variety of other liver diseases such as (i) nonalcoholic steatohepatitis (NASH), which occurs in patients with type II diabetes and obesity, after intestinal bypass surgery, and can be induce by certain drugs, (ii) chronic cholestasis, particularly primary biliary cirrhosis, (iii) copper intoxication as it occurs in patients with Wilson disease and Indian childhood cirrhosis, (iv) hepatocellular carcinoma[24,25] (Fig. 1).

The common occurrence of alterations of the hepatocytic keratin cytoskeleton in such a variety of diseases poses the question whether there is a unique pathogenetic principle leading to this phenotype. Although the above mentioned diseases originate from different aetiologies, increased oxidative stress is a constant and common feature. For example, in ASH oxidation of ethanol results in the formation acetaldehyde, which depletes glutathione and disturbs mitochondrial function, both of which result in increased oxidative stress. Furthermore metabolism of alcohol by cytochrome P450 2E1 leads to generation of reactive oxygen species (ROS) as by-product.[26] In NASH oxidative stress is an indirect consequence of mitochondrial overload with free fatty acids, and the increased release of TNF-α from adipose tissue.[27] In chronic cholestasis there is accumulation of bile acids in hepatocytes, which also results in oxidative stress either directly by affecting signaling and mitochondrial function or indirectly by activation of polymorphonuclear granulocytes.[28] In copper toxicosis copper leads to oxidative stress by mediating the formation of hydroxyl radicals by the Fenton reaction. The biological role of oxidative stress in copper storage-associated liver diseases was recently underlined by the common presence of MBs containing the stress-induced, ubiquitin binding protein p62 in livers with Wilson disease or Indian childhood cirrhosis.[29] In liver cancer ballooned tumor cells containing MBs or cytoplasmic hyaline bodies, which share features with MBs and can be considered as precursors of MBs, are found in 10-20% of cancer cases[30] (Denk et al, manuscript submitted). Comparative gene expression profiling of HCC with and without inclusions demonstrated increased expression of heat shock proteins, p62 and keratin in HCC containing inclusions, which supports the hypothesis that formation of cytoplasmic

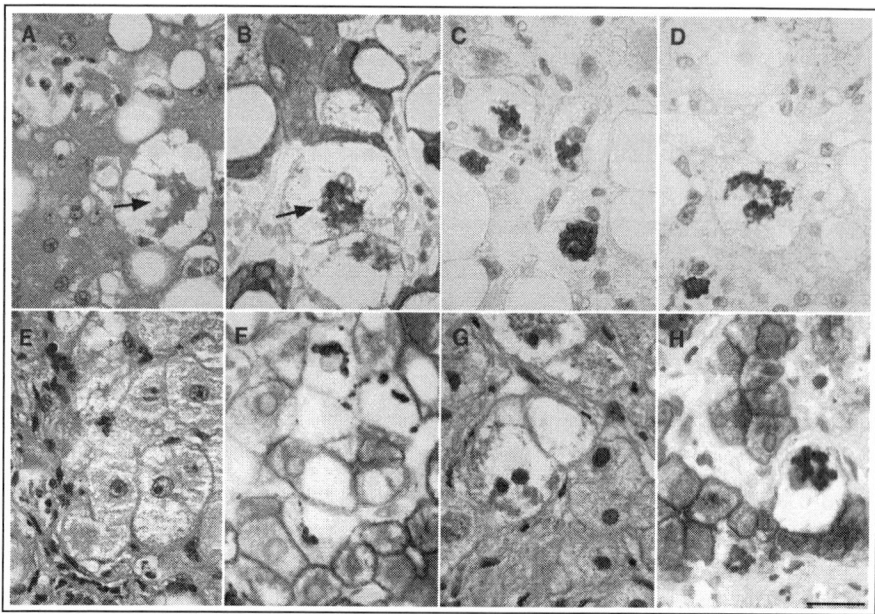

Figure 1. Alterations of the keratin cytoskeleton in human liver diseases. A) Hematoxylin eosin-stained human liver section with alcoholic hepatitis. B) Same liver as shown in (A) stained with antibodies to K8 and 18, C) stained with an antibody to p62, D) stained with an antibody to ubiquitin. Arrows in (A,B) indicate ballooned hepatocytes with reduced keratin intermediate filament network and Mallory bodies. E) Hematoxylin eosin-stained human liver section with nonalcoholic steatohepatitis. F) Same liver as shown in (E) stained with antibodies to K8 and 18. G) Hematoxylin eosin-stained human liver with Wilson disease. H) Same liver as shown in (G) stained with antibodies to K8 and 18. Scale bar = 30 μm.

inclusions is associated with increased stress (Neumann et al, unpublished observations). This observation is in line with the emerging general evidence that cancer cells experience increased oxidative stress and reveal altered responses to reactive oxygen species.[31]

Animal Models to Study Hepatocytic Keratin Alterations

The alterations of the keratin cytoskeleton observed in human liver diseases can be reproduced in mice by feeding a diet containing griseofulvin (GF) or 3,5-diethoxycarbonyl-1,4-dihydrocollidine (DDC)[24,32,33] (Fig. 2). Time course studies in GF- or DDC-fed mice, immunohistochemical and biochemical analyses of MBs as well as studies in K8 and K18 knock-out mice provided insights into the functional relevance of the keratin alterations in liver disease. DDC intoxication rapidly (within 2 days) leads to phosphorylation of keratin IFs at various phosphorylation sites.[34] Some of these phosphorylation sites are targets for protein kinases typically activated in stress conditions indicating that DDC induces a stress response in hepatocytes.[35] Furthermore mice respond to intoxication with marked overexpression of K8 and K18 resulting in a denser keratin IF network in enlarged hepatocytes[34,36] (Fig. 3). This overexpression of keratin apparently is an active (defence)response of hepatocytes to toxic injury since mice with one deleted K8 allele (K8+/-), which had limited capacities to increase keratin expression, were more sensitive to toxic injury than wild-type mice.[37] The concept that keratin overexpression confers resistance to drug toxicity is not restricted to DDC intoxication since in a previous report marked induction of multidrug resistance in cultured mouse fibroblasts was achieved by transfection with K8 and K18 cDNAs.[38] Interestingly, the overexpression of keratin in hepatocytes is a specific reaction to certain types of injury, such as ethanol, DDC, GF or

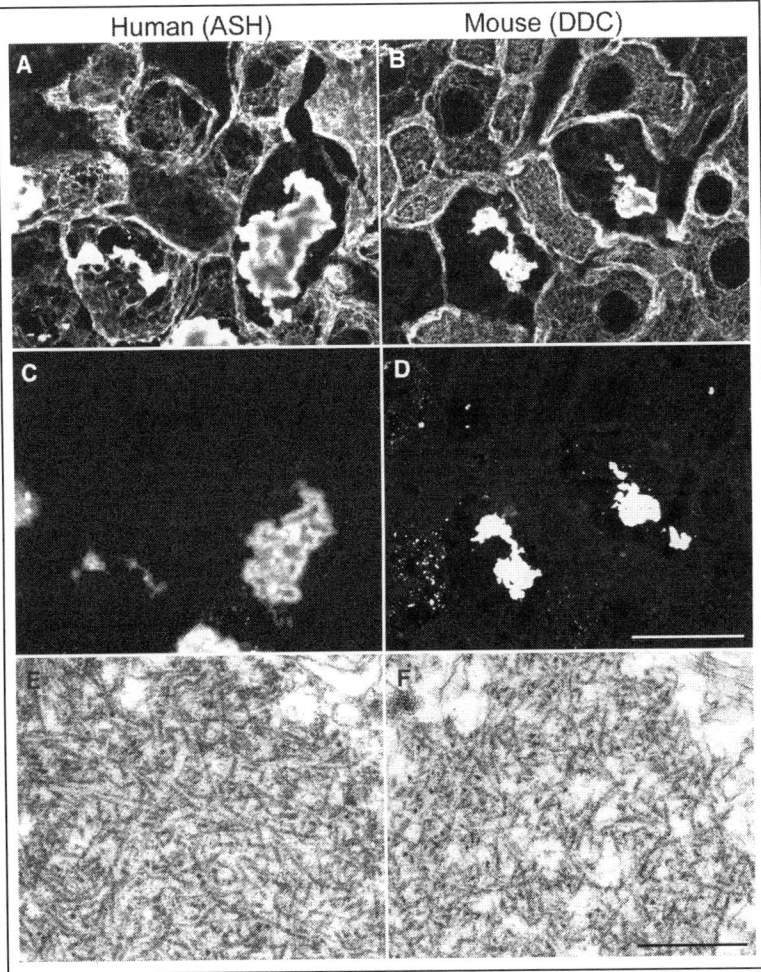

Figure 2. Double-label immunofluorescence microscopy of human liver with alcoholic steatohepatitis (ASH) (A,C) and mouse liver treated with DDC for 2 months (B,D) using antibodies to K8 and 18 (A,B) and to p62 (C,D). Electron microscopy of a Mallory body in human alcoholic steatohepatitis (E) and DDC-treated mouse liver (F). Note that alterations of hepatocytes in DDC-treated mice resemble human alcoholic steatohepatitis (i.e., loss of keratin intermediate filaments in ballooned hepatocytes and formation of Mallory bodies consisting of aggregated filamentous proteins). Scale bar = 20 μm (A-D); Scale bar = 250 nm (E,F).

cholestasis and not seen after intoxication with a series of cytostatic drugs[7,34,36,39] (Lackner et al, unpublished observation). The mode of transcriptional control, which leads to these rather specific alterations in keratin expression in liver diseases, is unknown. Continuation of DDC or GF intoxication for up to two months leads to a reduction of the density to a loss of cytoplasmic keratin IFs in hepatocytes (Fig. 3). The loosening of the keratin network is accompanied by the appearance small MBs at intersections of the keratin IF network at around 4-6 weeks of DDC intoxication. After 8 weeks larger MBs are found typically in the perinuclear region without association with keratin IFs (Fig. 3). These alterations are reversible upon cessation of intoxication. After 4 weeks of recovery only granular remnants of MBs are

Figure 3. Time course of keratin alterations in DDC-treated mice. Immunofluorescence microscopy using antibodies to K8 and 18 of normal mouse liver (A), after 7 days DDC feeding (B), after 2 months DDC feeding (C), after 2 months DDC feeding and 1 month of recovery from intoxication (D), and after 3 days of DDC refeeding of recovered mice (E). Note that after 1 month of recovery there are several hepatocytes with no detectable keratin filaments (empty hepatocytes), which still contain remnants of Mallory bodies at the cell periphery (arrow heads in d). Scale bar = 20μm.

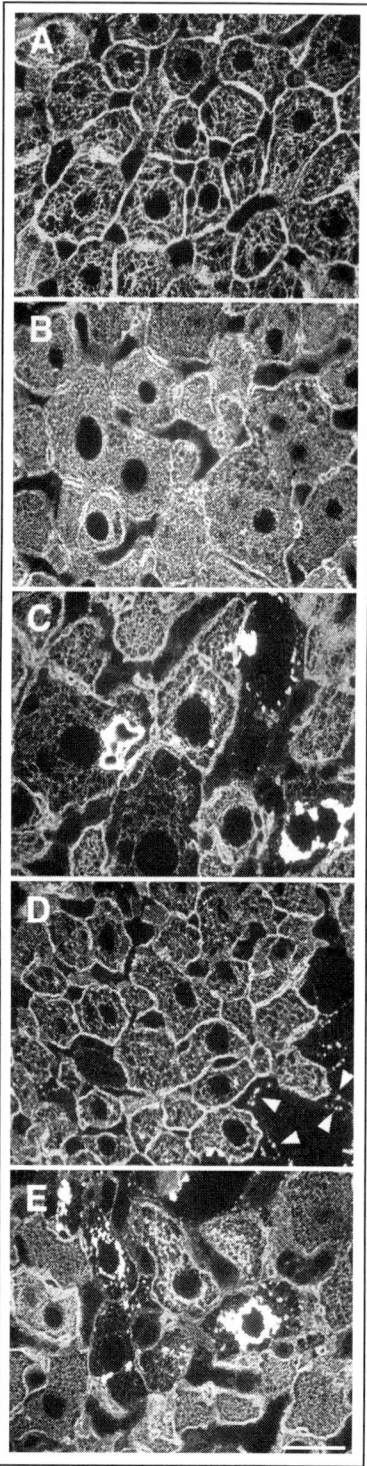

found at the cell periphery preferentially in association with desmosomes of hepatocytes most of which are still devoid of a cytoplasmic keratin IF network (Fig. 3). At this stage of recovery mice rapidly respond with MB formation upon reexposure to DDC.[34] This enhanced reappearance of MBs is also seen in humans with ASH if patients are exposed to alcohol even after prolonged periods of abstinence and is reminiscent of a "toxic memory response". Recent comparative studies in naïve mice and mice recovered from DDC intoxication, which were exposed to DDC for 3 days revealed marked differences in cytochrome P450 contents in these two groups of mice (Stumptner et al, unpublished observation). These findings indicate a relationship between DDC metabolism and cytoskeletal alterations. DDC is known to be metabolized by cytochrome P450 by N-demethylation, which leads to the generation of a methyl radical.[40] This indicates that chronic radical injury is a major consequence of DDC intoxication and the putative common pathogenetic principle in human ASH, NASH and other diseases associated with MB formation.

Mallory Bodies as a Product of the Cellular Response to Misfolded Keratin

Occurrence of misfolded proteins is a major consequence of radical injury. Oxidation of amino acid residues results in conformational changes and exposure of hydrophobic residues at the protein surface. Such misfolded proteins are inactive, can disturb cellular functions and have the tendency to aggregate via hydrophobic interactions.[41] Several cellular defence mechanisms against misfolded proteins exist. Misfolded proteins can either be refolded under consumption of ATP by heat shock protein 70 (HSP), preserved in a folding competent state by HSP 25/27, tagged by ubiquitin for degradation by the proteasome or aggregate in association with ubiquitin binding protein p62 as

Table 1. Summary of major Mallory body components

Mallory Body Component	Reference
Keratin	Denk et al, 1979[23]
	Franke et al, 1979[51]
Ubiquitin	Lowe et al, 1988[52]
	Ohta et al, 1988[54]
Ubiquitin	Lowe et al, 1988[52]
	Ohta et al, 1988[54]
p62	Zatloukal et al, 2002[50]
HSP70	Zatloukal et al, 2002[50]
HSP25	Zatloukal et al, 2002[50]
αB-crystallin	Lowe et al, 1992[53]

cytoplasmic inclusions recently designated in cultured cells as aggresomes or sequestosomes.[41-50] Analysis of the protein composition of MBs by mass spectrometry revealed that MBs consist predominantly of keratin, ubiquitinated keratin, HSP 70, HSP 25 and p62.[50] Immunohistochemical studies which confirmed the mass spectrometry data revealed in addition the presence α-B crystallin, another member of the small HSP family, and proteasomal subunits[23,24,51-55] (Table 1). The protein composition of MBs indicates hat in these diseases all major cellular defence mechanism against misfolded proteins (i.e., misfolded keratin) are involved. Abnormal conformation of keratins in human ASH and experimentally induced MBs has been demonstrated by monoclonal antibodies recognizing conformation-dependent epitopes, the excess of keratin 8 over keratin 18 in MBs, which is not compatible with proper assembly of IFs, and the reduction in α-helically structured domains.[34,56-58] These data imply that keratin is the primary target for modification by radicals and misfolding, whereas all other MB components reflect the cellular response to misfolded proteins. The central role of keratins in MB formation has been demonstrated with keratin 8 and keratin 18 null mice. Keratin 18 null mice responded to DDC-intoxication with enhanced MB formation and formed MBs spontaneously at high age[4] (Stumptner et al, manuscript submitted). In contrast no MBs were formed in the absence of keratin 8[37] (Fig. 4). Surprisingly, keratin 8 null mice, which did not form MBs revealed markedly increased toxicity, whereas keratin 18 null mice tolerated DDC intoxication like wild type mice. This discrepancy in the phenotypes of DDC-treated keratin 8 null and keratin 18 null mice allows drawing several important conclusions: (i) Keratin 8 is the core component of MBs and all other MB components bind to or coassemble with keratin, (ii) MBs by themselves are not toxic to cells but MB occurrence is associated with better tolerance to toxic stress, (iii) since in both types of knock-out mice hepatocytes are devoid of a keratin IF cytoskeleton, the increased sensitivity of keratin 8 null mice cannot by related to the loss of keratin IFs but indicates that keratins fulfil in addition nonstructural roles in toxic stress situations.[37,59]

How Can Keratins Influence Toxic Cell Injury?

Keratins were shown to undergo extensive modifications in various cell culture and in vivo conditions. They include phosphorylation, ubiquitination, proteolytic cleavage, and cross linking, which modulate the status of the keratin IF cytoskeleton and the interaction of keratins with several cellular proteins, particularly signaling molecules.[35,54,60,61] K8 and 18 are substrates of a variety of protein kinases involved in mitosis, apoptosis and stress.[35] Antibodies directed against specific phosphoepitopes on keratin revealed that K8 and 18 become phosphorylated at many sites in human ASH as well as in DDC-fed mice.[62] Some of these phosphorylation sites were targets of stress-induced protein kinases and were phosphorylated already within one

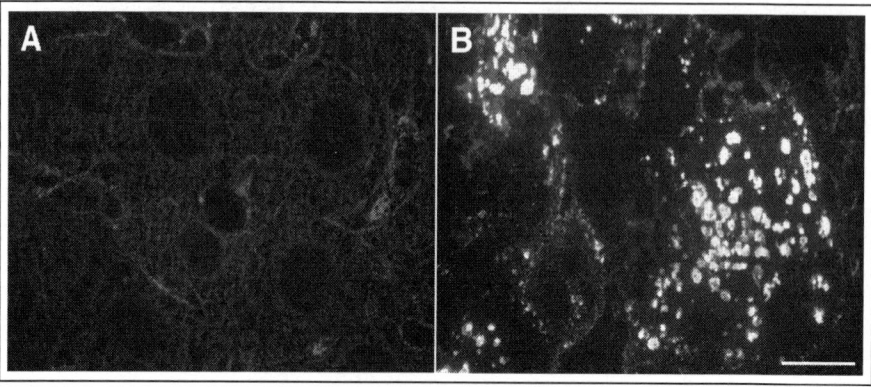

Figure 4. Requirement of K8 for Mallory body formation. Immunofluorescence microscopy of mouse livers stained with antibodies to K8 and K18. A) K8 null mouse liver treated for 2 months with DDC. B) K18 null mouse liver at 18 month of age. Note that in the absence of K8 no Mallory bodies are formed after DDC treatment. In contrast K18 null mice spontaneously formed numerous Mallory bodies at high age. Scale bar = 10 μm.

day of intoxication. This indicates that phosphorylation of keratin might be one of the initial mechanisms of involving keratins in toxic stress responses. A direct role of keratin phosphorylation as modulator of cell toxicity was demonstrated in transgenic mice expressing a K18 with a mutated phosphorylation site. These mice were more sensitive to treatment with griseofulvin or microcystin than wild-type mice or transgenic mice expressing nonmutated K18.[63] The functional relationship between keratin phosphorylation and toxicity could be based on the fact that phosphorylation status of keratins regulates IF assembly/disassembly as well as interaction of keratin with various signaling proteins. For example, a phosphorylation-dependent interaction of K18 with 14-3-3 was demonstrated recently.[64] 14-3-3 occurs in various isoforms and is an important integrator of major signaling pathways regulating cell proliferation and apoptosis.[65] The biological significance of the interaction of keratin with 14-3-3 was show in mice with an impaired keratin cytoskeleton, in which an association between alterations of keratin, 14-3-3 and cell proliferation was observed.[66] Furthermore, keratin is able to modulate TNF-α signaling. Association of keratin to tumor necrosis factor receptor 2 (TNFR2) influences TNF-α-induced activation of Jun NH_2-terminal kinase (JNK) as well as NFκB.[67] The interaction of keratin with TNF-α signaling could play a central role in ASH and NASH where enhanced production of TNF-α and activation of JNK and NFκB are constant findings.[27,68-70] Furthermore keratins were shown to modulate induction and execution of apoptosis by interfering at various levels with the apoptotic program. In K8-deficient mice there is enhanced targeting of Fas to the plasma membrane, which results in three- to four-fold increased sensitivity to Fas-mediated apoptosis.[71] A similar observation of increased Fas-mediated apoptosis was made in mice expressing mutated K18.[72] Furthermore K8- as well as in K18- deficient mice were up to 100-fold more sensitive to TNF-α-induced cell death, which was explained by the interaction of K8 and K18 with TNFR2.[67] Other studies, however, did only demonstrate a relationship between K8 or K18 and Fas but not TNF-α- induced apoptosis.[71,72] A further mechanism by which keratin modulates Fas or TNF-α mediated apoptosis was recently demonstrated in keratin 8 null mice. These mice showed marked reduction of c-Flip and did not respond with Erk1/2 activation upon stimulation of Fas, TNF-α receptor or TNF-α-related apoptosis-inducing ligand receptor.[73] A direct role of keratin in this pathway was suggested by the fact that keratin was found in association with a complex consisting of c-Flip, Raf, and Erk1/2.

Keratins are also targets in and modulators of the execution of apoptosis since keratins are cleaved by and associate with caspases. Cleavage of keratins by caspases leads to a disruption of the keratin IF network and the formation of small globular cytoplasmic inclusions containing cleaved and hyperphosphorylated keratins as well as activated caspases.[74] The association of activated caspases with keratin inclusions is a characteristic feature of epithelial cell apoptosis and is expected to influence the activity of caspases in the execution of the apoptotic program.

Mutations of Keratin Genes and Liver Diseases

Heterozygous mutations in the K8 and K18 genes were found in 17 out of 467 (3.6%) explanted livers from patients with end-stages of various acute and chronic liver diseases.[75-77] 75% of all mutations identified were at K8 Y53H, K8 G61C and K18 H127L, suggesting existence of mutation hot spots. The keratin mutations were shown to lead to reorganization and partial collapse of the keratin IF network in patient livers.[77] Furthermore the mutations affected keratin assembly in transfected cells if they were exposed to oxidative stress.[76] The preference of keratin mutations in patients with cryptogenic liver cirrhosis indicates that the presence of a keratin mutation might predispose to the development of liver cirrhosis, in particular in patients with NASH since NASH is regarded as the major cause of cryptogenic liver cirrhosis.

Since keratin mutations were found also in a healthy cohort in a frequency of 0.6%, heterozygous keratin mutations per se do not lead to liver disease but could rather act as a genetic modifier. Interestingly, keratin mutations were not found in patients with earlier stages of liver diseases.[78] This discrepancy to the studies performed by B. Omary in end-stage liver disease[75-77] underlines that heterozygous mutations of K8 and K18 are not the cause of liver disease in humans but could characterize patients with risk for enhanced progression of liver disease to cirrhosis requiring liver transplantation.

References

1. Franke WW, Denk H, Kalt R et al. Biochemical and immunological identification of cytokeratin proteins present in hepatocytes of mammalian liver tissue. Exp Cell Res 1981; 131:299-318.
2. Moll R, Franke WW, Schiller D et al. The catalog of human cytokeratins: Patterns of expression in normal epithelia, tumors and cultured cells. Cell 1982; 31:11-24.
3. Fuchs E, Weber K. Intermediate filaments: Structure, dynamics, function, and disease. Annu Rev Biochem 1994; 63:345-382.
4. Magin TM, Schröder R, Leitgeb S et al. Lessons from keratin 18 knockout mice; formation of novel keratin filaments, secondary loss of keratin 7 and accumulation of liver specific keratin 8-positive aggregates. J Cell Biol 1998; 140:1441-1451.
5. Van Eyken P, Sciot R, Desmet VL. A cytokeratin immunohistochemical study of alcoholic liver disease: Evidence that hepatocytes can express "bile duct-type" cytokeratins. Histopathology 1988; 13:605-617.
6. Van Eyken P, Sciot R, Desmet VL. A cytokeratin immunohistochemical study of cholestatic liver disease: Evidence that hepatocytes can express "bile duct-type" cytokeratins. Histopathology 1989; 15:125-135.
7. Fickert P, Trauner M, Fuchsbichler A et al. Cytokeratins as targets for bile acid induced toxicity. Am J Pathol 2002; 160:491-499.
8. Van Eyken P, Sciot R, Callea F et al. The development of the intrahepatic bile ducts in man: A keratin-immunohistochemical study. Hepatology 1988; 8:1586-1595.
9. Roskams TA, Libbrecht L, Desmet VJ. Progenitor cells in diseased human liver. Semin Liver Dis 2003; 23:385-396.
10. Baribault H, Price H, Miyai K et al. Mid-gestational lethality in mice lacking keratin 8. Genes Dev 1993; 7:1191-1202.
11. Baribault H, Penner J, Iozzo RV et al. Colorectal hyperplasia and inflammation in keratin 8-deficient FVB/N mice. Genes Dev 1994; 8:2964-2974.
12. Hesse M, Franz T, Tamai Y et al. Targeted deletion of keratins 18 and 19 leads to trophoblast fragility and early embryonic lethality. EMBO J 2000; 19:5060-5070.
13. Omary BM, Ku NO, Toivola DM. Keratins: Guardians of the liver. Hepatology 2002; 35:251-257.

14. Herrmann H, Hesse M, Reichenzeller M et al. Functional complexity of intermediate filament cytoskeletons: From structure to assembly to gene ablation. Int Rev Cytol 2003; 223:83-175.
15. Fuchs E. Epidermal differentiation: The bare essentials. J Cell Biol 1990; 111:2807-2814.
16. Paladini RD, Takahashi K, Bravo NS et al. Onset of reepithelialization after skin injury correlates with a reorganization of keratin filaments in wound edge keratinocytes: Defining a potential role for keratin 16. J Cell Biol 1996; 132:381-397.
17. McGowan K, Coulombe PA. The wound repair-associated keratins 6, 16, and 17. Insightsinto the role of intermediate filaments in specifying keratinocyte cytoarchitecture. Subcell Biochem 1998; 31:173-204.
18. Paramio JM, Casanova ML, Segrelles C et al. Modulation of cell proliferation by cytokeratins K10 and K16. Mol Cell Biol 1999; 19:3086-3094.
19. Paramio JM, Segrelles C, Ruiz S et al. Inhibition of protein kinase B (PKB) and PKCzeta mediates keratin K10-induced cell cycle arrest. Mol Cell Biol 2001; 21:7449-7459.
20. Burt AD, Mutton A, Day CP. Diagnosis and interpretation of steatosis and steatohepatitis. Sem Diagnostic Pathol 1998; 15:246-258.
21. Hall P. Pathological spectrum of alcoholic liver disease. In: Hall P, ed. Pathology and Pathogenesis: Alcoholic Liver Disease. London, Boston, Melbourne, Auckland: Edward Arnold, 1995:41-68.
22. Brunt E. Nonalcoholic Steatohepatitis. Semin Liver Dis 2004; 24:3-20.
23. Denk H, Franke WW, Eckerstorfer R et al. Formation and involution of Mallory bodies (alcoholic hyalin) in murine and human liver revealed by immunofluorescence microscopy with antibodies to prekeratin. Proc Natl Acad Sci USA 1979; 76:4112-4116.
24. Denk H, Stumptner C, Zatloukal K. Mallory body revisited. J Hepatol 2000; 32:689-702.
25. Zatloukal K, Stumptner C, Fuchsbichler A et al. The keratin cytoskeleton in liver disease. J Pathol 2004; 204:367-376.
26. Lieber CS. Alcoholic liver disease: New insights in pathogenesis lead to new treatments. J Hepatol 2000; 32:113-128.
27. Angulo P. Nonalcoholic fatty liver disease. N Engl J Med 2002; 16:1221-1231.
28. Fang Y, Han SI, Mitchell C et al. Bile acids induce mitochondrial ROS, which promote activation of receptor tyrosine kinases and signaling pathways in rat hepatocytes. Hepatology 2004; 40:961-971.
29. Muller T, Langner C, Fuchsbichler A et al. Immunohistochemical analysis of Mallory bodies in Wilsonian and non hepatic copper toxicosis. Hepatology 2004; 39:963-969.
30. Stumptner C, Heid H, Fuchsbichler A et al. Analysis of intracytoplasmic hyaline bodies in a hepatocellular carcinoma. Demonstration of p62 as major constituent. Am J Pathol 1999; 154:1701-1710.
31. Benhar M, Engelberg D, Levitzki A. ROS, stress activated kinases and stress signaling in cancer. EMBO Rep 2002; 3:420-425.
32. Denk H, Gschnait F, Wolff K. Hepatocellular hyalin (Mallory bodies) in long term griseofulvin-treated mice: A new experimental model for the study of hyalin formation. Lab Invest 1975; 32:773-776.
33. Tsunoo C, Harwood TR, Arak S et al. Cytoskeletal alterations leading to Mallory body formation in livers of mice fed 3,5-diethoxycarbonyl-1,4-dihydrocollidine. J Hepatol 1987; 5:85-97.
34. Stumptner C, Fuchsbichler A, Lehner K et al. Sequence of events in the assembly of Mallory body components in mouse liver: Clues to the pathogenesis and significance of Mallory body formation. J Hepatol 2001; 34:665-675.
35. Omary MB, Ku NO, Liao J et al. Keratin modifications and solubility properties in epithelial cells and in vitro. Subcell Biochem 1998; 31:105-140.
36. Cadrin M, Hovington H, Marceau N et al. Early perturbation in keratin and actin gene expression and fibrillar organisation in griseofulvin-fed mouse liver. J Hepatol 2000; 33:199-207.
37. Zatloukal K, Stumptner C, Lehner M et al. Cytokeratin 8 protects from hepatotoxicity, and its ratio to cytokeratin 18 determines the ability of hepatocytes to form Mallory bodies. Am J Pathol 2000; 156:1263-1274.
38. Bauman PA, Dalton WS, Anderson JM et al. Expression of cytokeratin confers multiple drug resistance. Proc Natl Acad Sci USA 1994; 91:5311-5314.
39. Fickert P, Trauner M, Fuchsbichler A et al. Bile acid-induced Mallory body formation in drug-primed mouse liver. Am J Pathol 2002; 161:2019-2026.
40. Tephly TR, Coffman BL, Ingall G et al. Identification of N-methylprotoporphyrin IX in livers of untreated mice and mice treated with 3,5-diethoxycarbonyl-1,4-dihydrocollidine: Source of the methyl group. Arch Biochem Biophys 1981; 212:120-126.
41. Grune T, Reinheckel T, Davies KJA. Degradation of oxidized proteins in mammalian cells. FASEB J 1997; 11:526-534.
42. Ehrnsperger M, Gräber S, Gaeste M et al. Binding of nonnative protein to Hsp25 during heat shock creates a reservoir of folding intermediates for reactivation. EMBO J 1997; 16:221-229.

43. Hartl FU. Molecular chaperones in cellular protein folding. Nature 1996; 381:571-580.
44. Vadlamudi RK, Joung I, Strominger J et al. p62, a phosphotyrosine-independent ligand of the SH2 domain of p56[lck], belongs to a new class of ubiquitin-binding proteins. J Bio Chem 1996; 271:20235-20237.
45. Shin J. P62 and the sequestosome, a novel mechanism for protein metabolism. Arch Pharm 1998; 21:629-633.
46. Wickner S, Maurizi MR, Gottesman S. Posttranslational quality control: Folding, refolding, and degrading proteins. Science 1999; 286:1888-1893.
47. Johnston JA, Ward CL, Kapito RR. Aggresomes: A cellular response to misfolded proteins. J Cell Biol 1998; 143:1883-1898.
48. Stumptner C, Fuchsbichler A, Heid H et al. Mallory body—A disease associated type of sequestosome. Hepatology 2002; 35:1053-1062.
49. Ciechanover A, Schwartz AL. Ubiquitin-mediated degradation of cellular proteins in health and disease. Hepatology 2002; 35:3-6.
50. Zatloukal K, Stumptner C, Fuchsbichler A et al. p62 a common component of the cellular response to aggregated proteins. Am J Pathol 2002; 160:255-263.
51. Franke WW, Denk H, Schmid E et al. Ultrastructural, biochemical and immunologic characterization of Mallory bodies in livers of griseofulvin-treated mice. Fimbriated rods of filaments containing prekeratin-like polypeptides. Lab Invest1979; 40:207-220.
52. Lowe J, Blanchard A, Morell K et al. Ubiquitin is a common factor in intermediate filament inclusion bodies of diverse type in man, including those of Parkinson's disease, Pick's disease, and Alzheimer's disease, as well as Rosenthal fibers in cerebellar astrocytomas, cytoplasmic bodies in muscle, and Mallory bodies in alcoholic liver disease. J Pathol 1988; 155:9-15.
53. Lowe J, McDermott H, Pike I et al. Alpha B crystallin expression in nonlenticular tissues and selective presence in ubiquitinated inclusion bodies in human disease. J Pathol 1992; 166:61-68.
54. Ohta M, Marceau N, Perry G et al. Ubiquitin is present on the cytokeratin intermediate filaments and Mallory bodies of hepatocytes. Lab Invest 1988; 59:848-856.
55. Riley NE, Li J, McPhaul LW et al. Heat shock proteins are present in Mallory bodies (cytokeratin aggresomes) in human liver biopsy specimens. Exp Mol Pathol 2003; 74:168-172.
56. Hazan R, Denk H, Franke WW et al. Change of cytokeratin organization during development of Mallory bodies as revealed by a monoclonal antibody. Lab Invest 1986; 54:543-553.
57. Zatloukal K, Fesus L, Denk H et al. High amount of ε - (γ-glutamyl) lysine cross-links in Mallory bodies. Lab Invest 1992; 66:774-777.
58. Cadrin M, French SW, Wong TT. Alteration in molecular structure of cytoskeleton proteins in griseofulvin-treated mouse liver: A pressure tuning infrared spectroscopy study. Exp Mol Pathol 1991; 55:170-179.
59. Coulombe PA, Omary MB. 'Hard' and 'soft' principles defining the structure, function and regulation of keratin intermediate filaments. Curr Opin Cell Biol 2002; 14:110-122.
60. Zatloukal K, Denk H, Lackinger E et al. Hepatocellular cytokeratins as substrates of transglutaminases. Lab Invest 1989; 61:603-608.
61. Ku N-O, Omary B. Keratins turn over by ubiquitination in a phosphorylation-modulated fashion. J Cell Biol 2000; 149:547-552.
62. Stumptner C, Omary MB, Fickert P et al. Hepatocyte cytokeratins are hyperphosphorylated at multiple sites in human alcoholic hepatitis and in a Mallory body mouse model. Am J Pathol 2000; 156:77-90.
63. Ku NO, Michie SA, Soetikno RM et al. Mutation of a major keratin phosphorylation site predisposes to hepatotoxic injury in transgenic mice. J Cell Biol 1998; 143:2023-2032.
64. Ku NO, Michie S, Resurreccion EZ et al. Keratin binding to 14-3-3 proteins modulates keratin filaments and hepatocyte mitotic progression. Proc Natl Acad Sci USA 2002; 99:4373-4378.
65. Hermeking H. The 14-3-3 cancer connection. Nat Rev Cancer 2003; 3:931-943.
66. Toivola DM, Nieminen MI, Hesse M et al. Disturbances in hepatic cell-cycle regulation in mice with assembly-deficient keratins 8/18. Hepatology 2001; 34:1174-1183.
67. Caulin C, Ware CF, Magin TM et al. Keratin-dependent, epithelial resistance to tumor necrosis factor-induced apoptosis. J Cell Biol 2000; 149:17-22.
68. Deaciuc IV. Alcohol and cytokine networks. Alcohol 1997; 14:421-430.
69. Tsukamoto H, Lu SC. Current concepts in the pathogenetic of alcoholic liver injury. FASEB J 2001; 15:1335-1349.
70. Diehl AM. Cytokine regulation of liver injury and repair. Immunol Rev 2000; 174:160-71.
71. Gilbert S, Loranger A, Daigle N et al. Simple epithelium keratins 8 and 18 provide resistance to Fas-mediated apoptosis. The protection occurs through a receptor- targeting modulation. J Cell Biol 2001; 154:763-773.

72. Ku NO, Soetikno RM, Omary MB. Keratin mutation in transgenic mice predisposes to fas but not TNF-Induced apoptosis and massive liver injury. Hepatology 2003; 37:1006-1014.
73. Gilbert S, Loranger A, Marceau N. Keratins modulate c-Flip/extracellular signal-regulated kinase 1 and 2 antiapoptotic signaling in simple epithelial cells. Mol Cell Biol 2004; 24:7072-7081.
74. MacFarlane M, Merrison W, Dinsdale D et al. Active caspases and cleaved cytokeratins are sequestered into cytoplasmic inclusions in TRAIL-induced apoptosis. J Cell Biol 2000; 148:1239-1254.
75. Ku NO, Wright TL, Terrault NA et al. Mutation of human keratin 18 in association with cryptogenic cirrhosis. J Clin Invest 1997; 99:19-23.
76. Ku NO, Gish R, Wright TL et al. Keratin 8 Mutations in patients with cryptogenic liver disease. N Engl J Med 2001; 344:1580-1587.
77. Ku NO, Darling JM, Krams SM et al. Keratin 8 and 18 mutations are risk factors for developing liver disease of multiple etiologies. PNAS 2003; 100:6063-6068.
78. Hesse M, Berg T, Wiedenmann B et al. A frequent keratin 8 p.L227L polymorphism, but no point mutations in keratin 8 and 18 genes, in patients with various liver disorders. J Med Gen 2004; 41:e42.

CHAPTER 10

The Search for Specific Keratin Functions:
The Case of Keratin K10

Mirentxu Santos, Carmen Segrelles, Sergio Ruiz, M. Fernanda Lara
and Jesús M. Paramio*

Abstract

The main function of the epidermis is to provide an essential barrier between the individual and the environment. This primarily stems from a finely regulated process of differentiation occurring in this stratified epithelium. Keratins are the most abundant proteins in this tissue and provide resistance against mechanical stress to the epidermal cells. Moreover, characteristic changes in the pattern of expression of this family of proteins takes place during the process of differentiation in epidermis. The K5/K14 pair, characteristic of basal proliferative cells, is switched off at the differentiation onset, and the expression of keratin pair K1/K10 concomitantly starts. In addition, under several pathological or physiological hyperproliferative situations K1/K10 pair expression is down-regulated. These facts have led to the assumption that specific functions can be exerted by each specific keratin polypeptide. Here we will summarize the recent and past effort to elucidate these particular roles. Collectively, the data have indicated particular roles for keratin K10, as a possible modulator of epidermal homeostasis contributing to the modulation of several signaling pathways in keratinocytes. The possibility that these functions are probably not unique among the large family of keratins will open new exciting and interesting scientific fields.

Introduction

Why Focusing on Keratin K10?

Most mammals live in an adverse environment: the terrestrial medium. As a consequence, they have developed an specialized organ to preserve their bodies and isolate from physical trauma: the skin. The protective functions of this structure, one of the biggest organs of the body, mostly reside in its epithelial component, the epidermis, and stems from a finely regulated program of differentiation. Cell differentiation, in general, is a fundamental process that provides selective identity through the specific gene expression programs in each tissue. The ample variety of different tissues in the organism represents the necessity of maintaining distinct structures to carry out specific functions.

The epidermis is a stratified epithelium whose major cell type is the keratinocyte. The program of differentiation in epidermis takes place when keratinocytes located in the innermost basal cell layer move to the upper strata until the formation of anucleated squames, which exert the protective function and are periodically shed and replaced.[1,2] This process requires and is characterized by changes in the biochemical properties of the cells that affect several

*Corresponding Author: Jesús M. Paramio—Department of Molecular and Cell Biology, CIEMAT, Madrid, Spain. Email: jesusm.paramio@ciemat.es

Intermediate Filaments, edited by Jesus Paramio. ©2006 Landes Bioscience and Springer Science+Business Media.

functions and that are finely tuned. Among these changes, one of the earliest events is the change in keratin expression pattern.[1,3,4] The keratinocytes in the basal layer mainly express keratins K5 and K14, however when these keratinocytes are committed into the differentiation program, down-regulate the expression of certain integrins and move upwards, they cease the expression of K5 and K14 genes, and switch on the expression of keratin K1 and K10 genes.[5] On the other hand, under situations associated with increased proliferation such as wound healing or in certain pathologies such as psoriasis or cancer, the expression of keratins K1 and K10 is decreased, and other keratins, which are not normally expressed in the interfollicular epidermis, such as K6 or K16, are then induced.[6-9] The functional reasons and the molecular events that might justify these changes remain obscure, however they imply that the properties of the cells change as they vary their characteristic keratin expression pattern, and would suggest that different keratin pairs may exert different functions in epidermal keratinocytes. This question is still a matter of debate, but clearly have allowed many research groups to investigate this problem using a variety of different approaches.

Beginning in the early nineties the analysis of certain transgenic mice and the genetic analysis of human hereditary syndromes unequivocally demonstrated that epidermal keratins provide cells with mechanical resilience against physical stress (reviewed in refs. 10-15). Nonetheless, if this is the single function, why nature have generated so many different proteins? And, why is their expression regulated in such a finely modulated and complex pattern?. The answer to these questions might be simply: different keratin polypeptides may have subtle structural differences so they can fulfill cell type-specific protective roles (for a detailed discussion see refs. 16,17) As a nonexclusive explanation, they might also have other cell type-specific functions. If so, what can be these specific roles? During the recent past years, our group, among others, has been studying this possibility with different methodological approaches focusing on the functional reasons for K10 expression.

Keratin K10 Assembly and Dynamics

In vivo, formation of K1/K10 intermediate filament network occurs in the cytoplasm of cells with a preexisting cytoskeleton composed mainly by keratins K5 and K14. This fact raised the question whether K1 and K10 might have specific requirements to form such cytoskeletal network. In pursuing the answer to this problem, Kartasova et al[18,19] performed a series of elegant experiments. They found that upon transfection in fibroblasts, the keratins K1 and K10 were unable to form an extensive keratin filament network by themselves, and only small isolated dense K1/K10 filament bundles were observed by electron microscopy.[18] This is in contrast with the findings of transfecting K5 and K14, or K8 and K18.[18] Moreover, once fibroblasts were expressing K5 and K14, the K1/K10 pair integrated normally without disruption of the cytoskeletal architecture.[18] The same authors also monitored if the expression of K1/K14 and K5/K10 pairs in fibroblasts gives rise to any architectural alteration, and found that the simply expression of K10 prevented the formation of a well developed cytoskeletal framework.[18] These results, similar to the findings of injecting epidermal mRNA,[20] are in contrast with the in vitro findings that led to the qualification of keratins as promiscuous proteins, as any type I keratin polypeptide can form intermediate filaments with any type II polypeptide.[21] Similar observations were performed upon permanent transfection of bovine K1 and K10 genes in bovine mammary gland cells, in which the decreased expression of endogenous keratins prevented the formation of well developed IF cytoskeleton, but rather short twisted filaments and small aggregates were produced.[22] Later on, we confirmed that K1/K10 expression, in contrast to K8/K18, was unable to form a keratin meshwork in fibroblasts (Fig. 1) and that K10 was able to integrate into the endogenous preexisting keratin cytoskeleton in transfected epithelial cells.[23] However, when we analyzed K10 expression at early times after transfection we found that regularly sized round aggregates were formed, and interacted with the endogenous keratin cytoskeleton. However, other cytoskeletal systems were also affected. Indeed, K10 aggregates promoted the disorganization and colocalization with vimentin, actin and tubulin (Fig. 2). On the other hand, these structures seemed to be

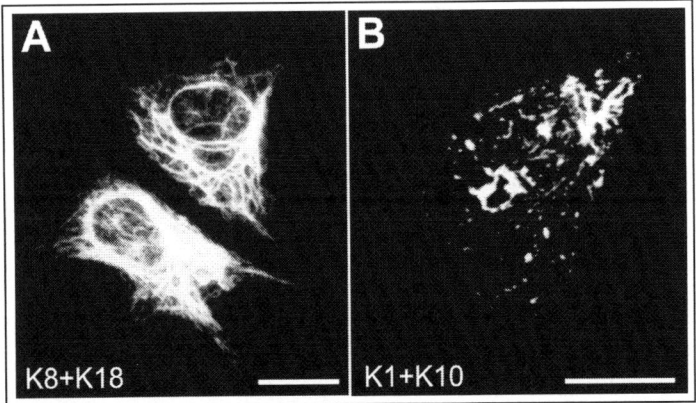

Figure 1. Keratins1 and 10 require a preexisting scaffold to form a well developed keratin cytoskeleton. Primary bovine fibroblasts were transfected with keratins K8 and K18 (A) or K1 and K10. Transfected proteins were detected by immunofluorescence using specific antibodies. Note that simple keratins form a normal cytoskeleton whereas K1 and K10 (B) form abnormal short, twisted filaments and aggregates. Bars = 10 μm.

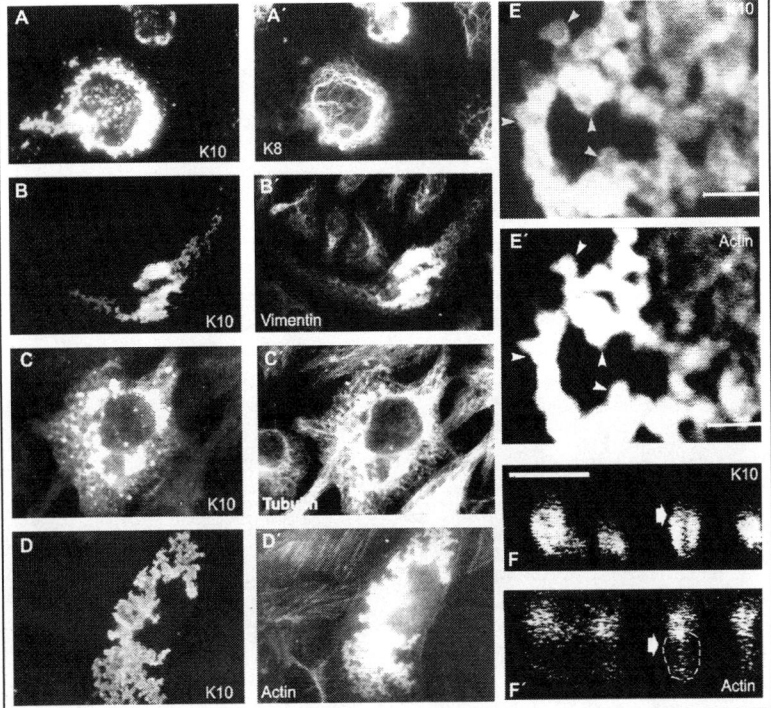

Figure 2. Keratin K10 aggregates interact with other cytoskeletal elements. Double immunofluorescence images of epithelial PtK2 cells transfected with human K10 gene (A, B, C, D) showing also the distribution of endogenous K8 Keratin (A'), vimentin (B'), tubulin (C') and actin (D'). E, E' Confocal images of keratin K10 aggregates (E) and actin (E'). Z-reconstruction showed that keratin K10 aggregates form hollow barrels (F) whereas actin appeared on top and penetrates inside these barrels (F'). Bars = 5 μm.

transitory and, without requiring protein synthesis, evolved toward the integration of K10 into the endogenous keratin cytoskeleton through a complex and highly dynamics process leading to a situation with all the cytoskeletal systems organized in a manner similar to that observed in nontransfected cells.[23]

The above commented results clearly indicated that keratins K1 and K10 have surprisingly dynamic properties different from those reported previously for other keratins such as K8 and K18.[24,25] However, this assembly dynamics can only be monitored and seems to be exclusive of cultured cells, as in K5-deficient mice[26] and in humans suffering from complete absence of K14,[27,28] although basal keratinocytes are devoid of keratin filaments, K1 and K10 were present and showed a distribution similar to that in control epidermis in suprabasal layers. In addition, these assembly properties can be exclusive to transfection experiments. In this regard, the dynamics of keratin assembly may differ between preexisting and newly synthesized keratins.[29] To analyze these aspects, a new set of experiments was designed. Two different types of epithelial cells were fused and the behavior of the parental keratin cytoskeletons were followed by double immunofluorescence at different times after heterokaryon formation.[30] In addition, this system allows using different drugs to monitor the effect of different cell processes in the dynamics of keratin copolymerization and assembly.[31] We observed that the process as a whole is very rapid, thus demonstrating the highly dynamic nature of the keratin cytoskeleton, and the involvement of protein phosphorylation.[30,31] However, significant differences were observed among the different polypeptides studied. In this regard K5, K10, K8 and K18 are mobilized and reassemble rapidly into the hybrid cytoskeleton (3-6 hours) even in the absence of protein synthesis, whereas K14 requires a substantially longer period (9-24 hours).[30] Interestingly, this effect was observed even when K5, K14 and K10 were initially in the same parental cytoskeleton (Fig. 3), thus demonstrating that different keratin polypeptides, even belonging to the same subfamily and being highly related, as is the case for K10 and K14, may display different dynamic properties in cells.

Keratin K10 Inhibits Cell Proliferation

During the above commented experiments, we noted the difficulty in obtaining permanent transfectants expressing K1 and/or K10. This observation could be in the line of severe cell perturbation by the formation of keratin aggregates reported previously.[18,19] However, as commented above, we found that K10 was integrated into the endogenous keratin cytoskeleton, and such aggregates were present in the cells only few hours after transfection,[23] thus making highly improbable that these structures were responsible for the perturbed cell survival. On the other hand, this might indicate that the expression of K10 interferes with cell proliferation.

To study this possibility, a series of transfection experiments were carried out in human keratinocytes. We observed that expression of K10, but not other keratins such as K13, K14 or K16, reduced proliferation of the transfected cells.[32] The case of K16 was also relevant as its expression seemed to accelerate proliferation, even under low serum conditions, and specifically reversed the K10-induced cell growth arrest.[32] If this observation is not an artifact, one might expect that the mechanism underlying this growth arrest might be related to any of the known mechanisms of physiological inhibition of proliferation. In fact, a number of several experimental data (Fig. 4). indicated that this seemed to be the actual situation and involved retinoblastoma (*Rb*) gene product as: (I) K10 inhibited entry into S-phase of the cell cycle; (II) the K10-induced inhibition was hampered by cotransfection with viral oncoproteins that interfere with pRb, but not with p53; (III) coexpression of specific cyclins, cyclin dependent kinases (CDK) or cyclin /CDK complexes abolished the K10-induced growth arrest; (IV) the expression of K10 reduced cyclin D1 expression and, concomitantly pRb phosphorylation.[32] In addition, K10 does not exert antiproliferative effects in Rb-deficient cells, but restored expression of *Rb* in these cells makes them again susceptible to the K10 inhibition of cell cycle progression.[32] More recently, others and we also observed similar situation in vivo. In epidermis lacking *Rb*, obtained through the generation of tissue specific knock out, K10/expressing cells proliferate.[33,34]

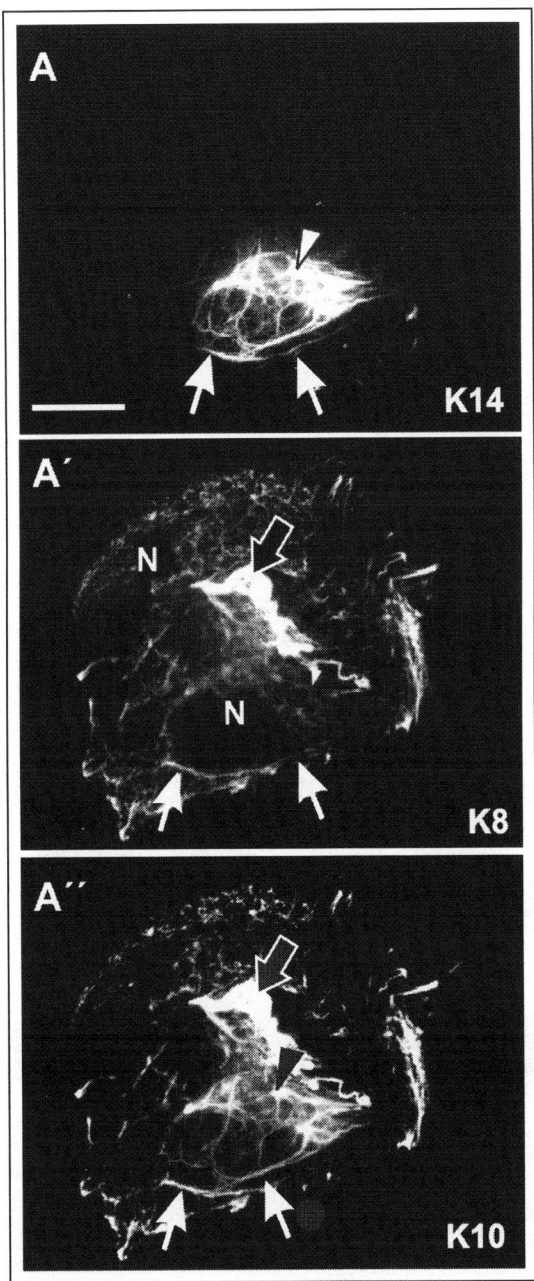

Figure 3. Different keratin polypeptides display distinct dynamic properties in living cells. Triple immunofluorescence against K14, K8 and K10 in PtK2-BMGE+HK10 heterokaryons three hours after cell fusion. Note that K8 and K10-containing filaments spread throughout the heterokaryon cytoplasm, whereas K14 is concentrated around the parental BMGE+Hk10 cell nucleus. White arrows denote K8+,K14+,K10+ filaments. Hollow arrows denote K8+, K10+, K14- filaments. Arrowheads denote the K14 filaments around parental BMGE+Hk10 cell nucleus. N denote parental nuclei. Bar = 10 μm.

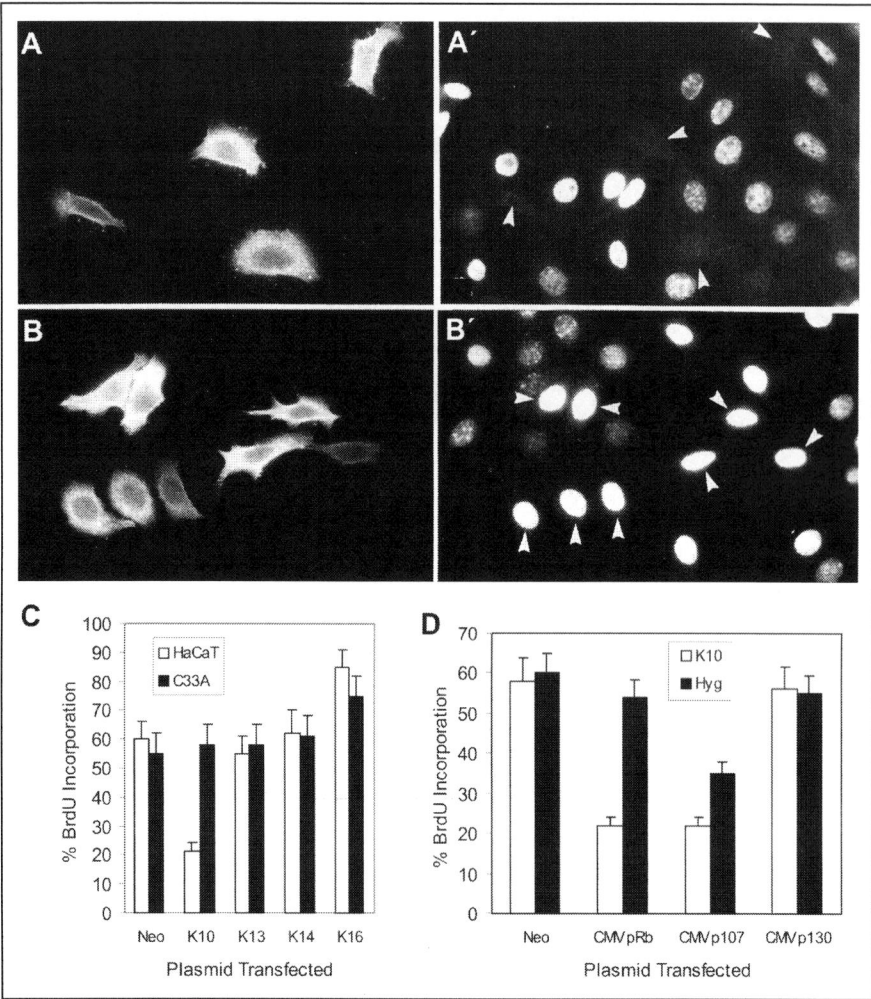

Figure 4. K10 expression inhibits cell proliferation in cultured cells. MCA3D murine keratinocytes were transiently transfected with K10 (A) or tagged K14 (B) and the expression of transfected keratins and bromodeoxyuridine incorporation (A', B') were monitored by double immunofluorescence. Transfected cells are denoted by arrowheads. Note the reduced bromodeoxyuridine incorporation in K10, but not in K14 expressing cells. C) Summary of five to ten independent experiments as those shown in A and B using the quoted keratin constructs. At least 500 transfected cells were scored in each experiment and BrdU was normalized to that scored in the nontransfected cells on each experiment. D) Summary of five independent experiments as those shown in C but using Rb-deficient C33A cells upon transfection with K10 (black bars) or with K10 plus the quoted pocket protein (white bars).

These data represent the first evidence of keratin functions other than merely structural. However, how can reconcile cell cycle progression with a cytoplasmic protein? There might be a missed link. We thus carried out similar rescue experiments using several effectors of the ras signaling pathway. This pathway was specifically selected because its involvement in mouse skin tumorigenesis[35] and its functional connection with pRb.[36,37] We observed that coexpression of PDK1, Akt and PKCζ effectively abolished the K10-induced cell cycle arrest, whereas raf, rhoA, Rac1, cdc42 and permanently active phophoinositide 3 kinase (PI3K) were unsuccessful

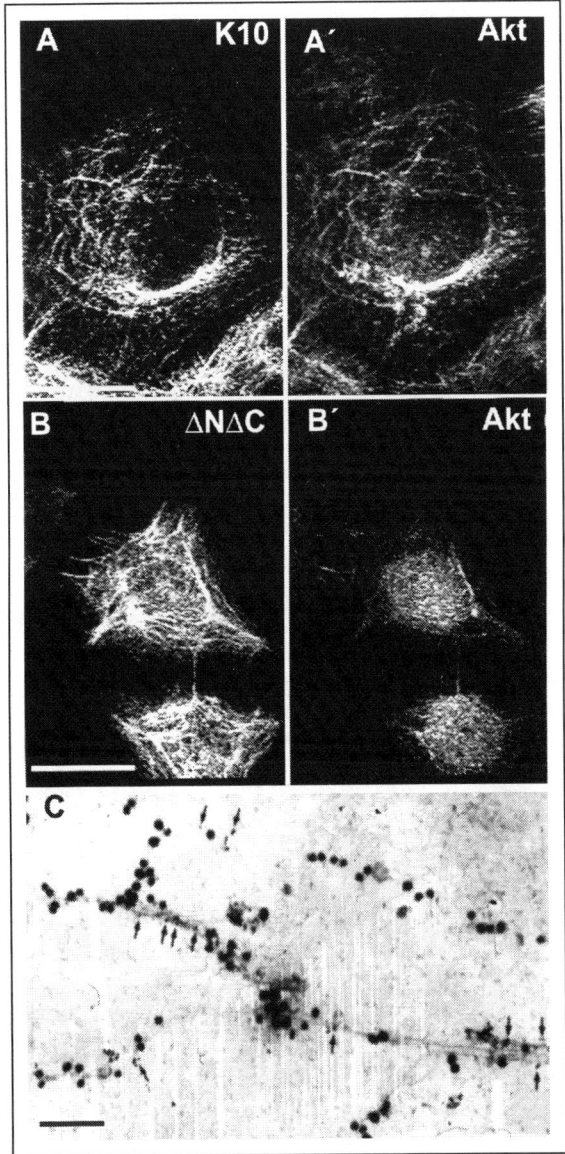

Figure 5. K10 interacts with Akt. Confocal images of HaCaT keratinocytes transfected with K10 (A) or a mutant K10 (ΔNΔC) lacking amino and carboxy termini (B) showing the distribution of the transfected protein and endogenous Akt (A', B'). Note that Akt binds to K10 but not to the ΔNΔC mutant. C) Immuno electron microscopy confirming that Akt (small arrows; 5 nm gold particles) binds to K10 containing filaments (10 nm gold particles).

in rescuing the K10-induced effects.[38] This implicates that K10 was specifically acting somehow downstream of PI3K, as the elements active in rescuing were effectors of this pathway. Indeed the mechanism seemed to involve the physical interaction between K10 and Akt and/or PKCζ as the wt K10 but not a mutant K10, lacking head and tail domains and inactive as cell cycle inhibitor,[32] interacted and colocalized with these two kinases (Fig. 5). This lead to the sequestration of the kinases at the IF cytoskeleton, precluding their translocation and subsequent activation,[38] and as a consequence, the expression of cyclin D1 and the phosphorylation of pRb are prevented.[38] It is worth mentioning that Akt exerts part of its oncogenic effects through the modulation of cyclin D1 levels and/or localization.[39-41]

Collectively, the above described results strongly indicated that keratin K10 is involved in the modulation of specific signal transduction pathways leading to the control of cell proliferation. These data are in agreement with its characteristic expression pattern in nonproliferative cells,[5] and are also in accordance with the premature loss of K10 expression found in many hyperproliferative diseases, and in particular skin tumors.[42] However, the relevance of the data is limited due to the fact of using cultured cells, and it might occur, as in the case of K10 assembly, that this system and the in vivo situation are different. Therefore further in vivo confirmation should be necessary.

Transgenic Mice Models to Study K10 Functions

At present three different classes of transgenic mice involving K10 gene have been generated, besides those using K10 regulatory elements to direct specific genes to the suprabasal layer of epidermis.[43,44] They are those used to analyze the consequences of K10 mutations and their relationship with epithelial fragility syndromes, those in which K10 expression is targeted ectopically to other cell types, and knock out models.

In one of the earliest models Fuchs' group demonstrated that the expression of a mutant K10 in transgenic mice[45] caused phenotypic alterations similar to those found in the epidermolytic hyperkeratosis (EHK), also known as bullous ichtyosiform erythroderma (BCIE) (OMIM #113800). Later on, the same group found mutations in K10 in BCIE patients.[46] Other transgenic mice models were also generated recapitulating many of the clinical features of this disorder, including both neonatal and adult forms of the disease.[47] In a clear and conclusive fashion it was demonstrated that mutations in K10, and also in K1, gene are causal of this syndrome and that K10 is required for the maintenance of both cellular and tissue integrity.[48-50] Therefore a common function between K10 and other epidermal keratins exists. However, some differences were observed in these models compared to those of mutations in K5 and/or K14. They included the increase of cell proliferation promoted in the basal layer keratinocytes and also that perturbations in the keratin IF network caused changes in the nuclear shape and aberrations in cell cytokinesis.[45]

The first attempt to ectopically express K10 in transgenic mice was reported by Blessing et al,[51] Using the rat insulin promoter sequences they directed K1 and K10 expression to the pancreatic islets β cells. They found that the synthesis of K10 was compatible with cell function. On the contrary, mice expressing K1 alone or in combination with K10 developed a special form of diabetes characterized by a drastic reduction of insulin secretory vesicles and also in insulin producing cells.[51] These alterations were attributed to the presence of abnormal keratin aggregates inside the nuclei of these β cells.[51] Of note, K10 was irrelevant in this system. This seems to be in disagreement with our cultured cell experiments, however, it is important to remark the proliferative rate of the β cells in vivo is extremely low. On the other hand, similar findings were initially obtained when K10 was expressed under the K6β promoter, as no phenotypic alterations were produced.[52] However, when these mice were used in skin carcinogenesis experiments the picture dramatically changed. The transgenic animals displayed a delayed tumor development.[52] This might indicate that K10 expression really intereferes with tumor development in vivo. However, the response was only partial probably due to the fact that the promoter used in these experiments, bovine K6β, is expressed in vivo only in cells with limited proliferative potential,[53,54] or due to the intrinsic susceptibility of the original mouse strain used. Finally, the expression level observed can not be sufficient to generate more drastic effects. Indeed, we observed in cultured cells using an inducible promoter that a threshold in the K10 expression level is necessary to promote the cell cycle arrest.[32] Because of these reasons, we generated another transgenic mouse model in which the expression of K10 is targeted to the basal compartment of the epidermis using the bK5 promoter.[55] The different transgenic mice founders obtained fall into two separate classes, those without any obvious phenotype and those displaying several abnormalities leading to premature death between days 60 and 90 after birth. Interestingly the major difference between these two classes was the number of copies of the transgene, which is also related to its expression level.[55,56] The early lethality displayed by the phenotypic mice precluded the establishment of lines, however, we observed that in the nonphenotypic mice when we brought the transgene to homozigosity the complete phenotype appeared, including the lethality in parallel with an increased expression of the K10.[57]

These mice provide new important insights into the K10 possible functions.[57] We observed a severe epidermal hypoplasia characterized by a reduced number of cells in the interfollicular epidermis. Indeed, the two-three layers of living cells observed in control mice (Fig. 6A). is reduced to a single layer, which also adopt a large flattened morphology reminiscent of some short of endothelial cells (Fig. 6A'). In addition, a clear hyperkeratotic phenotype was evident (Fig. 6A'). This phenotype is displayed throughout adult epidermis, but some regions also display a normal appearance. Remarkably, the absence of histological phenotype in these areas was coincident with ectopic expression of K6.[57] The electron microscopy analyses showed that the epidermal cells displayed abnormally flattened nuclei (Fig. 6B') compared with control (Fig. 6B). However, normal bundles of filaments (arrows in Fig. 6B, B', B") and desmosomes (hollow arrows in Fig. 6B, B', B") were similarly observed in transgenic and non transgenic cells. Another features that characterized the transgenic samples were the presence of degenerative mitochondria and a clear hallo, devoid of most cytoplasmic organelles, around the nuclei of the cells (h in Fig. 6B"). These are classic characteristic of cells in the early stages of apoptotic processes. The molecular characterization of the phenotype demonstrated reduced proliferation (Fig. 6C) in parallel with reduced phosphorylation of pRb and reduced activation of Akt and PKCζ kinases (Fig. 6D).[57] These results and those performed previously with bK6βhK10 transgenic mice,[52] prompted us to analyze the tumor susceptibility in bK5hK10 transgenic mice. However, given the early lethality of the phenotypic mice we were forced to restrict our analyses to those mice bearing low copies of the transgene. When we bred these mice to TG.AC and applied TPA, we observed a severe reduction in the number of the tumors compared to control TG.AC mice (Fig. 6E), moreover the tumors were smaller and displayed a lower capacity to undergo malignant transformation.[57] Collectively the data obtained in this model allowed us to establish that K10 expression can induce, in a dose dependent manner, the proliferation arrest of the cells associated with cell cycle and specific signal transduction pathway inhibition in a cell autonomous mode. Interestingly, the reduced tumor susceptibility also pointed to the fact that K10 may act as a tumor suppressor in an in vivo context, and highlights the relevance of Akt in the two stage mouse skin carcinogenesis.[58]

The Paradigms of K10 Knock Out Model

Perhaps the most appropriate approach for elucidating protein functions involves inactivation of the corresponding gene followed by analysis of the resulting phenotype. So far, two different models of K10 knock out mice have been generated. In a first attempt, Porter et al[59] targeted the K10 gene and found a different phenotype in the homozygotes and heterozygotes, although both of them exhibited similarities to BCIE. Homozygotes suffered from severe skin fragility and died shortly after birth. Heterozygotes were apparently unaffected at birth, but developed hyperkeratosis with age.[59] In both genotypes, aggregation of keratin intermediate filaments, changes in keratin expression, and alterations in epidermal differentiation were observed.[59] The fact that a severe phenotype also occurred in heterozygosis pointed to a dominant effect, which is not a common feature of knock out models. In fact, careful analyses of the phenotype at molecular level revealed that hetero- and homozygotes expressed a truncated keratin 10 peptide.[60] This is, besides other different alterations involving the specific interaction with restricted subsets of keratins, the most probably cause of the epidermolytic hyperkeratosis observed, in close resemblance with other transgenic mice expressing mutant K10 constructs.[46,47]

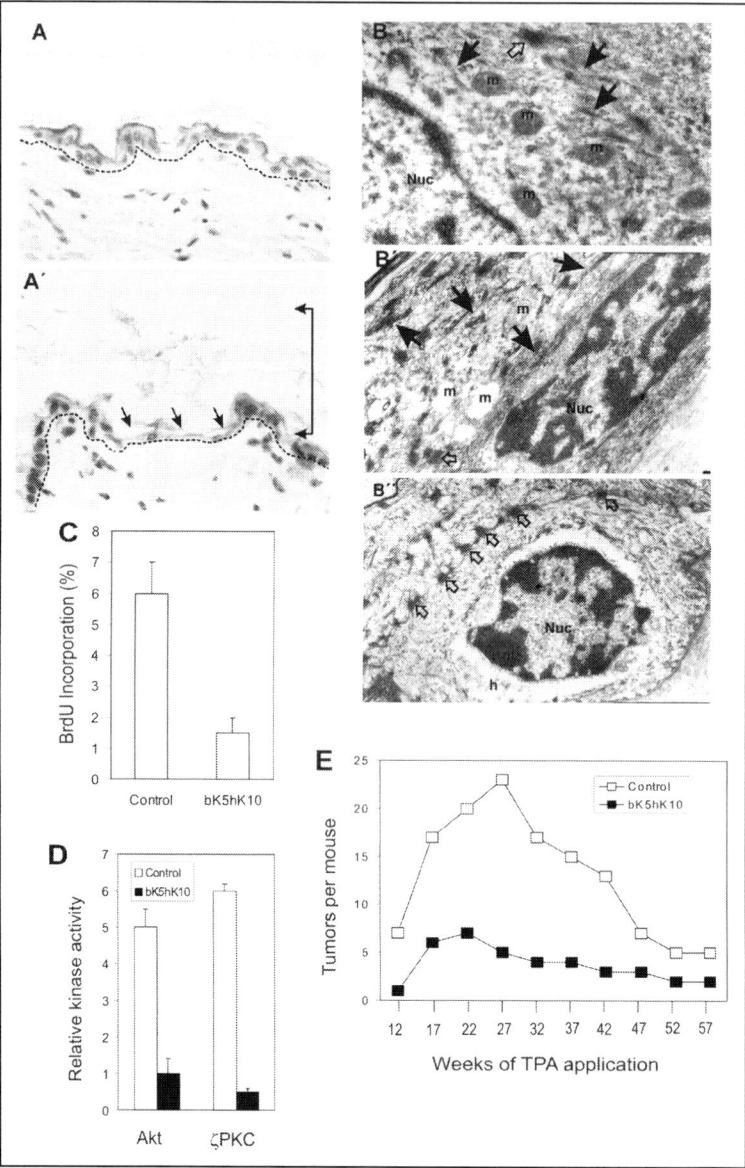

Figure 6. Abnormalities due K10 ectopic expression in transgenic mice. In contrast to control mice (A), the bK5hK10 transgenic mice (A') display hypoplasic epidermis with flattened cells(arrows) and hyperkeratotic appearance (bracket). Dashed line denotes dermal-epidermal junction. Electron microscopy showed that these flattened cells of bK5hK10 mice (B') have normal filament bundles (arrows) similar to those observed in control mice (B). In addition normal desmosomes are formed (hollow arrows in B"). Of note is that mitochondria (denoted by m) display a degenerative appearance in transgenic mouse epidermal cells. Also a clear hallo (h) around the nuclei of transgenic mouse epidermal cells is evident in some cases (B"). C) Summary of BrdU incorporation in epidermal cells of 30 days old mice. D) Quantitative analyses of Akt and PKCζ activities in extracts from control (white bars) and transgenic bK5hK10 mice (black bars) epidermis. E) Tumor multiplicity in Control TG.AC (open squares) and double TG.AC bK5hK10 transgenic mice. Note reduction in bK5hK10 background.

More recently, the same group generated a "true" K10 knock out mice.[61] They observed that a normal epidermis without signs of fragility is formed in K10 null mice. As in many other models, the very mild phenotype was associated to functional redundancy mediated by the sustained expression of basal keratins in the suprabasal layers.[61] On the other hand, striking alterations were observed when knockout mice reached the adult age connecting cell proliferation and K10 expression.[62] Indeed, adult K10-null mice epidermis showed hyperproliferation of basal keratinocytes accompanied by the induction of c-Myc, cyclin D1, 14-3-3σ and ectopic expression of keratins K6 and K16.[62] Surprisingly, the phosphorylation status of Rb remained unaltered, in contrast with the increased proliferation and cycD1 expression found, which would indicate increased phosphorylation of Rb.[62] While these data suggest that K10 not only serves structural functions, they also imply that K10 is able to alter the proliferation state of basal cells, thus suggesting that K10 can also act in a non cell autonomous manner.[62] More recently, these authors have also performed two stage carcinogenesis in K10-null mice (Reichelt et al, in press). They found a reduced formation of papilloma due to accelerated epidermal cell turnover, although the tumor incidence was similar in control and knock out mice. Moreover, increased response to TPA both in basal and suprabasal layers was demonstrated in K10-null epidermis, which in agreement showed increased ERK and p38 activation (Reichelt et al, in press). These results are indicative that K10 is not necessary for the expression of suprabasal proliferation. However, there are several unanswered questions that would require further analyses. Whereas increased c-myc and cycD1 expression was found in null epidermis, in papillomas their expression was similar in control and K10-deficient samples. This indicates that these two gene are rapidly induced in controls, whilst in knock out either cannot be induced or they have reached their maximum expression. This last seems not very probable, thus indicating that carcinogenesis-induced expression is defective in null epidermis. Of note, it has been reported that, whereas increased cycD1 has no effect in tumor susceptibility,[63] increased c-myc expression is able to promote the formation of tumors in skin.[64] On the other hand, the observed activation of ERK, ectopic expression of integrin α_6 and reduced activin expression seem to be in contrast with the reduced tumor formation.

Reconciling Hypotheses

The fact that in the absence of K10 there is only moderate ectopic proliferation in the suprabasal layers of the epidermis[62] (Reichelt et al, in press) seems to be in contrast with our findings in vivo and in vitro indicating that K10 caused cell cycle arrest. However, we may consider again the possibility of functional redundancy. Indeed, there is ectopic expression of other keratins in the suprabasal layers of K10-null mice.[62] On the other hand, the possibility that keratin K10 might posses cell autonomous and noncell autonomous functions is highly attractive.

We have analyzed how other pathways are affected in the bK5hK10 transgenic mice. Given that Akt is somehow upstream the NFκB pathway, and this is highly relevant in epidermis[65-70] we initially focused in this aspect.[71] As expected, the inhibition of Akt promoted by K10 led to decreased basal NFκB activity, and primary keratinocytes derived from the bK5hK10 mice displayed a poor response to interleukin 1β and TNFα two well-known activators of the NFκB cascade.[71] While biochemical studies revealed that Akt might modulate the levels of the β and γ subunits of the IKK complex,[71] thus opening an interesting link, our studies also provided an unexpected findings. We observed increased production of interleukin 1 and 6 and TNFα in the epidermis of the bK5hK10 mice.[71] This unexpected production of chemoattractant chemokines might explain the acute inflammatory response found when K10 is expressed to a high levels in the suprabasal layers of the oral mucosa[72] through the recruitment of lymphocytes. Moreover, this cytokines also have important effects in epidermal cells.[73-76] Interestingly, TNFα has been shown to promote the inhibition of keratinocyte proliferation.[74,77,78] Therefore, it is possible to speculate that in K10-null mice there might be reduced interleukin production leading to increased basal cell proliferation. The possibility that altered cytokine production might also alter carcinogenesis susceptibility should also be considered.

Other aspect to take into account is the absence of increased proliferation in the suprabasal layers of the K10-null mice. In this regard, it should be extremely simplistic to consider that K10 is the major modulator of cell proliferation of differentiating keratinocytes. Indeed, the coupling between proliferation and differentiation is a highly relevant issue that affect many cell circuits lying at the end in the intimate cell cycle control. In agreement, others and we have recently shown that the specific ablation of the retinoblastoma gene in epidermis leads to aberrant proliferation of differentiating cells.[33,34] One then have to consider that this mechanism should be still active in K10 null keratinocytes, which indeed show no alterations in Rb phosphorylation.[62] However, the loop between K10 and Rb requires further analysis as we observed reduced expression of K10 in Rb-deficient keratinocytes.[34] The generation of compound transgenic mice lacking Rb and K10 and/or ectopically expressing K10 might help to understand this possible complex issue.

Finally, we also can consider that the expression or absence of K10 may also affect other signaling pathways. In this regard we have recently found that in bK5hK10 mice the thymus shows a dramatic phenotype characterized by premature thymus involution due to K10 expression in the epithelial medullary cells (Santos et al, in press). Moreover, we characterized altered T cell proliferation, apoptosis and differentiation in close association with altered Notch signaling (Santos et al, in press). Interestingly, Notch expression also showed alterations in the epidermis of bKhK10 mice (Santos et al, in press). These data provide new evidence of the involvement of K10 in paracrine signaling. Given the reported relevance of Notch pathway in epidermal differentiation and carcinogenesis,[79-83] and the fact that K10 is coexpressed with Notch members in interfollicular epidermis,[84,85] this pathway should be considered as possible subject of analysis in K10-null mice.

Overall, the work done with K10 along the last years by several groups has revealed new exciting aspects of keratin Biology. Regardless the differences observed in several pathways between ectopic K10 or null transgenic mice, it is now evident that this protein may affect cell signaling in a cell autonomous and non cell autonomous manners, including paracrine and merocrine mechanisms. Still many aspects are obscure to get the whole picture of the molecular mechanisms acting in K10-expressing or null cells, but the possibility that these are not only associated with keratin K10, but with other proteins of this family as well, has really opened new visions for the future research with this amazing family of polypeptides.

Acknowledgements

Research is supported by grants from the Ministerio de Educación y Ciencia, Comunidad Autónoma de Madrid and Fundació La Caixa. Expert technical support by Pilar Hernández and Jesús Martínez-Palacio is specially acknowledged.

References

1. Fuchs E, Byrne C. The epidermis: Rising to the surface. Curr Opin Genet Dev 1994; 4(5):725-736.
2. Fuchs E, Raghavan S. Getting under the skin of epidermal morphogenesis. Nat Rev Genet 2002; 3(3):199-209.
3. Byrne C, Tainsky M, Fuchs E. Programming gene expression in developing epidermis. Development 1994; 120(9):2369-2383.
4. Kopan R, Fuchs E. A new look into an old problem: Keratins as tools to investigate determination, morphogenesis, and differentiation in skin. Genes Dev 1989; 3(1):1-15.
5. Fuchs E, Green H. Changes in keratin gene expression during terminal differentiation of the keratinocyte. Cell 1980; 19(4):1033-1042.
6. Weiss RA, Eichner R, Sun TT. Monoclonal antibody analysis of keratin expression in epidermal diseases: A 48- and 56-kdalton keratin as molecular markers for hyperproliferative keratinocytes. J Cell Biol 1984; 98(4):1397-1406.
7. Takahashi K, Coulombe PA. Defining a region of the human keratin 6a gene that confers inducible expression in stratified epithelia of transgenic mice. J Biol Chem 1997; 272(18):11979-11985.

8. Paladini RD, Takahashi K, Bravo NS et al. Onset of reepithelialization after skin injury correlates with a reorganization of keratin filaments in wound edge keratinocytes: Defining a potential role for keratin 16. J Cell Biol 1996; 132(3):381-397.

9. McGowan K, Coulombe PA. The wound repair-associated keratins 6, 16, and 17. Insights into the role of intermediate filaments in specifying keratinocyte cytoarchitecture. Subcell Biochem 1998; 31:173-204.

10. Fuchs E, Coulombe PA. Of mice and men: Genetic skin diseases of keratin. Cell 1992; 69(6):899-902.

11. Takahashi K, Coulombe PA, Miyachi Y. Using transgenic models to study the pathogenesis of keratin-based inherited skin diseases. J Dermatol Sci 1999; 21(2):73-95.

12. Irvine AD, McLean WH. Human keratin diseases: The increasing spectrum of disease and subtlety of the phenotype-genotype correlation. Br J Dermatol 1999; 140(5):815-828.

13. McLean WH, Lane EB. Intermediate filaments in disease. Curr Opin Cell Biol 1995; 7(1):118-125.

14. Lane EB. Keratin diseases. Curr Opin Genet Dev 1994; 4(3):412-418.

15. Porter RM, Lane EB. Phenotypes, genotypes and their contribution to understanding keratin function. Trends Genet 2003; 19(5):278-285.

16. Coulombe PA, Omary MB. 'Hard' and 'soft' principles defining the structure, function and regulation of keratin intermediate filaments. Curr Opin Cell Biol 2002; 14(1):110-122.

17. Paramio JM, Jorcano JL. Beyond structure: Do intermediate filaments modulate cell signalling? Bioessays 2002; 24(9):836-844.

18. Kartasova T, Roop DR, Holbrook KA et al. Mouse differentiation-specific keratins 1 and 10 require a preexisting keratin scaffold to form a filament network. J Cell Biol 1993; 120(5):1251-1261.

19. Kartasova T, Roop DR, Yuspa SH. Relationship between the expression of differentiation-specific keratins 1 and 10 and cell proliferation in epidermal tumors. Mol Carcinog 1992; 6(1):18-25.

20. Kreis TE, Geiger B, Schmid E et al. De novo synthesis and specific assembly of keratin filaments in nonepithelial cells after microinjection of mRNA for epidermal keratin. Cell 1983; 32(4):1125-1137.

21. Hatzfeld M, Franke WW. Pair formation and promiscuity of cytokeratins: Formation in vitro of heterotypic complexes and intermediate-sized filaments by homologous and heterologous recombinations of purified polypeptides. J Cell Biol 1985; 101(5 Pt 1):1826-1841.

22. Blessing M, Jorcano JL, Franke WW. Enhancer elements directing cell-type-specific expression of cytokeratin genes and changes of the epithelial cytoskeleton by transfections of hybrid cytokeratin genes. EMBO J 1989; 8(1):117-126.

23. Paramio JM, Jorcano JL. Assembly dynamics of epidermal keratins K1 and K10 in transfected cells. Exp Cell Res 1994; 215(2):319-331.

24. Kulesh DA, Oshima RG. Cloning of the human keratin 18 gene and its expression in nonepithelial mouse cells. Mol Cell Biol 1988; 8(4):1540-1550.

25. Kulesh DA, Cecena G, Darmon YM et al. Posttranslational regulation of keratins: Degradation of mouse and human keratins 18 and 8. Mol Cell Biol 1989; 9(4):1553-1565.

26. Peters B, Kirfel J, Bussow H et al. Complete cytolysis and neonatal lethality in keratin 5 knockout mice reveal its fundamental role in skin integrity and in epidermolysis bullosa simplex. Mol Biol Cell 2001; 12(6):1775-1789.

27. Rugg EL, McLean WH, Lane EB et al. A functional "knockout" of human keratin 14. Genes Dev 1994; 8(21):2563-2573.

28. Chan Y, Anton-Lamprecht I, Yu QC et al. A human keratin 14 "knockout": The absence of K14 leads to severe epidermolysis bullosa simplex and a function for an intermediate filament protein. Genes Dev 1994; 8(21):2574-2587.

29. Lu X, Quinlan RA, Steel JB et al. Network incorporation of intermediate filament molecules differs between preexisting and newly assembling filaments. Exp Cell Res 1993; 208(1):218-225.

30. Paramio JM, Casanova ML, Alonso A et al. Keratin intermediate filament dynamics in cell heterokaryons reveals diverse behaviour of different keratins. J Cell Sci 1997; 110(Pt 9):1099-1111.

31. Paramio JM. A role for phosphorylation in the dynamics of keratin intermediate filaments. Eur J Cell Biol 1999; 78(1):33-43.

32. Paramio JM, Casanova ML, Segrelles C et al. Modulation of cell proliferation by cytokeratins K10 and K16. Mol Cell Biol 1999; 19(4):3086-3094.

33. Balsitis SJ, Sage J, Duensing S et al. Recapitulation of the effects of the human papillomavirus Type 16 E7 oncogene on mouse epithelium by somatic Rb deletion and detection of pRb-independent effects of E7 in Vivo. Mol Cell Biol 2003; 23(24):9094-9103.

34. Ruiz S, Santos M, Segrelles C et al. Unique and overlapping functions of pRb and p107 in the control of proliferation and differentiation in epidermis. Development 2004; 131(11):2737-2748.
35. Quintanilla M, Brown K, Ramsden M et al. Carcinogen-specific mutation and amplification of Ha-ras during mouse skin carcinogenesis. Nature 1986; 322(6074):78-80.
36. Lee KY, Ladha MH, McMahon C et al. The retinoblastoma protein is linked to the activation of Ras. Mol Cell Biol 1999; 19(11):7724-7732.
37. Mittnacht S, Paterson H, Olson MF et al. Ras signalling is required for inactivation of the tumour suppressor pRb cell-cycle control protein. Curr Biol 1997; 7(3):219-221.
38. Paramio JM, Segrelles C, Ruiz S et al. Inhibition of protein kinase B (PKB) and PKCzeta mediates keratin K10-induced cell cycle arrest. Mol Cell Biol 2001; 21(21):7449-7459.
39. Diehl JA, Cheng M, Roussel MF et al. Glycogen synthase kinase-3beta regulates cyclin D1 proteolysis and subcellular localization. Genes Dev 1998; 12(22):3499-3511.
40. Basso AD, Solit DB, Munster PN et al. Ansamycin antibiotics inhibit Akt activation and cyclin D expression in breast cancer cells that overexpress HER2. Oncogene 2002; 21(8):1159-1166.
41. Leis H, Segrelles C, Ruiz S et al. Expression, localization, and activity of glycogen synthase kinase 3beta during mouse skin tumorigenesis. Mol Carcinog 2002; 35(4):180-185.
42. Roop DR, Krieg TM, Mehrel T et al. Transcriptional control of high molecular weight keratin gene expression in multistage mouse skin carcinogenesis. Cancer Res 1988; 48(11):3245-3252.
43. Blessing M, Schirmacher P, Kaiser S. Overexpression of bone morphogenetic protein-6 (BMP-6) in the epidermis of transgenic mice: Inhibition or stimulation of proliferation depending on the pattern of transgene expression and formation of psoriatic lesions. J Cell Biol 1996; 135(1):227-239.
44. Bailleul B, Surani MA, White S et al. Skin hyperkeratosis and papilloma formation in transgenic mice expressing a ras oncogene from a suprabasal keratin promoter. Cell 1990; 62(4):697-708.
45. Fuchs E, Esteves RA, Coulombe PA. Transgenic mice expressing a mutant keratin 10 gene reveal the likely genetic basis for epidermolytic hyperkeratosis. Proc Natl Acad Sci USA 1992; 89(15):6906-6910.
46. Cheng J, Syder AJ, Yu QC et al. The genetic basis of epidermolytic hyperkeratosis: A disorder of differentiation-specific epidermal keratin genes. Cell 1992; 70(5):811-819.
47. Bickenbach JR, Longley MA, Bundman DS et al. A transgenic mouse model that recapitulates the clinical features of both neonatal and adult forms of the skin disease epidermolytic hyperkeratosis. Differentiation 1996; 61(2):129-139.
48. Rothnagel JA, Fisher MP, Axtell SM et al. A mutational hot spot in keratin 10 (KRT 10) in patients with epidermolytic hyperkeratosis. Hum Mol Genet 1993; 2(12):2147-2150.
49. Rothnagel JA, Greenhalgh DA, Wang XJ et al. Transgenic models of skin diseases. Arch Dermatol 1993; 129(11):1430-1436.
50. Rothnagel JA, Dominey AM, Dempsey LD et al. Mutations in the rod domains of keratins 1 and 10 in epidermolytic hyperkeratosis. Science 1992; 257(5073):1128-1130.
51. Blessing M, Ruther U, Franke WW. Ectopic synthesis of epidermal cytokeratins in pancreatic islet cells of transgenic mice interferes with cytoskeletal order and insulin production. J Cell Biol 1993; 120(3):743-755.
52. Santos M, Ballestin C, Garcia-Martin R et al. Delays in malignant tumor development in transgenic mice by forced epidermal keratin 10 expression in mouse skin carcinomas. Mol Carcinog 1997; 20(1):3-9.
53. Ramirez A, Vidal M, Bravo A et al. Analysis of sequences controlling tissue-specific and hyperproliferation-related keratin 6 gene expression in transgenic mice. DNA Cell Biol 1998; 17(2):177-185.
54. Ramirez A, Vidal M, Bravo A et al. A 5'-upstream region of a bovine keratin 6 gene confers tissue-specific expression and hyperproliferation-related induction in transgenic mice. Proc Natl Acad Sci USA 1995; 92(11):4783-4787.
55. Ramirez A, Bravo A, Jorcano JL et al. Sequences 5' of the bovine keratin 5 gene direct tissue- and cell-type-specific expression of a lacZ gene in the adult and during development. Differentiation 1994; 58(1):53-64.
56. Ramirez A, Milot E, Ponsa I et al. Sequence and chromosomal context effects on variegated expression of keratin 5/lacZ constructs in stratified epithelia of transgenic mice. Genetics 2001; 158(1):341-350.
57. Santos M, Paramio JM, Bravo A et al. The expression of keratin k10 in the basal layer of the epidermis inhibits cell proliferation and prevents skin tumorigenesis. J Biol Chem 2002; 277(21):19122-19130.
58. Segrelles C, Ruiz S, Perez P et al. Functional roles of Akt signaling in mouse skin tumorigenesis. Oncogene 2002; 21(1):53-64.

59. Porter RM, Leitgeb S, Melton DW et al. Gene targeting at the mouse cytokeratin 10 locus: Severe skin fragility and changes of cytokeratin expression in the epidermis. J Cell Biol 1996; 132(5):925-936.
60. Reichelt J, Bauer C, Porter R et al. Out of balance: Consequences of a partial keratin 10 knock-out. J Cell Sci 1997; 110(Pt 18):2175-2186.
61. Reichelt J, Bussow H, Grund C et al. Formation of a normal epidermis supported by increased stability of keratins 5 and 14 in keratin 10 null mice. Mol Biol Cell 2001; 12(6):1557-1568.
62. Reichelt J, Magin TM. Hyperproliferation, induction of c-Myc and 14-3-3sigma, but no cell fragility in keratin-10-null mice. J Cell Sci 2002; 115(Pt 13):2639-2650.
63. Rodriguez-Puebla ML, LaCava M, Conti CJ. Cyclin D1 overexpression in mouse epidermis increases cyclin-dependent kinase activity and cell proliferation in vivo but does not affect skin tumor development. Cell Growth Differ 1999; 10(7):467-472.
64. Pelengaris S, Littlewood T, Khan M et al. Reversible activation of c-Myc in skin: Induction of a complex neoplastic phenotype by a single oncogenic lesion. Mol Cell 1999; 3(5):565-577.
65. Schmidt-Ullrich R, Aebischer T, Hulsken J et al. Requirement of NF-kappaB/Rel for the development of hair follicles and other epidermal appendices. Development 2001; 128(19):3843-3853.
66. Courtois G, Israel A. NF-kappa B defects in humans: The NEMO/incontinentia pigmenti connection. Sci STKE 2000; 2000(58):E1.
67. Hu Y, Baud V, Oga T et al. IKKalpha controls formation of the epidermis independently of NF-kappaB. Nature 2001; 410(6829):710-714.
68. Takeda K, Takeuchi O, Tsujimura T et al. Limb and skin abnormalities in mice lacking IKKalpha. Science 1999; 284(5412):313-316.
69. Bell S, Degitz K, Quirling M et al. Involvement of NF-kappaB signalling in skin physiology and disease. Cell Signal 2003; 15(1):1-7.
70. Kaufman CK, Fuchs E. It's got you covered. NF-kappaB in the epidermis. J Cell Biol 2000; 149(5):999-1004.
71. Santos M, Perez P, Segrelles C et al. Impaired NF-kappa B activation and increased production of tumor necrosis factor alpha in transgenic mice expressing keratin K10 in the basal layer of the epidermis. J Biol Chem 2003; 278(15):13422-13430.
72. Santos M, Bravo A, Lopez C et al. Severe abnormalities in the oral mucosa induced by suprabasal expression of epidermal keratin K10 in transgenic mice. J Biol Chem 2002; 277(38):35371-35377.
73. Komine M, Rao LS, Freedberg IM et al. Interleukin-1 induces transcription of keratin K6 in human epidermal keratinocytes. J Invest Dermatol 2001; 116(2):330-338.
74. Cheng J, Turksen K, Yu QC et al. Cachexia and graft-vs.-host-disease-type skin changes in keratin promoter-driven TNF alpha transgenic mice. Genes Dev 1992; 6(8):1444-1456.
75. Pasparakis M, Courtois G, Hafner M et al. TNF-mediated inflammatory skin disease in mice with epidermis-specific deletion of IKK2. Nature 2002; 417(6891):861-866.
76. Turksen K, Kupper T, Degenstein L et al. Interleukin 6: Insights to its function in skin by overexpression in transgenic mice. Proc Natl Acad Sci USA 1992; 89(11):5068-5072.
77. Basile JR, Zacny V, Munger K. The cytokines tumor necrosis factor-alpha (TNF-alpha) and TNF-related apoptosis-inducing ligand differentially modulate proliferation and apoptotic pathways in human keratinocytes expressing the human papillomavirus-16 E7 oncoprotein. J Biol Chem 2001; 276(25):22522-22528.
78. Basile JR, Eichten A, Zacny V et al. NF-kappaB-mediated induction of p21(Cip1/Waf1) by tumor necrosis factor alpha induces growth arrest and cytoprotection in normal human keratinocytes. Mol Cancer Res 2003; 1(4):262-270.
79. Nicolas M, Wolfer A, Raj K et al. Notch1 functions as a tumor suppressor in mouse skin. Nat Genet 2003; 33(3):416-421.
80. Yamamoto N, Tanigaki K, Han H et al. Notch/RBP-J signaling regulates epidermis/hair fate determination of hair follicular stem cells. Curr Biol 2003; 13(4):333-338.
81. Okuyama R, Nguyen BC, Talora C et al. High commitment of embryonic keratinocytes to terminal differentiation through a Notch1-caspase 3 regulatory mechanism. Dev Cell 2004; 6(4):551-562.
82. Thelu J, Rossio P, Favier B. Notch signalling is linked to epidermal cell differentiation level in basal cell carcinoma, psoriasis and wound healing. BMC Dermatol 2002; 2(1):7.
83. Nickoloff BJ, Qin JZ, Chaturvedi V et al. Jagged-1 mediated activation of notch signaling induces complete maturation of human keratinocytes through NF-kappaB and PPARgamma. Cell Death Differ 2002; 9(8):842-855.
84. Rangarajan A, Talora C, Okuyama R et al. Notch signaling is a direct determinant of keratinocyte growth arrest and entry into differentiation. EMBO J 2001; 20(13):3427-3436.
85. Lefort K, Dotto GP. Notch signaling in the integrated control of keratinocyte growth/differentiation and tumor suppression. Semin Cancer Biol 2004; 14(5):374-386.

Index